SpringerBriefs in Electrical and Computer Engineering

For further volumes:
http://www.springer.com/series/10059

Yabo Li

In-Phase and Quadrature Imbalance

Modeling, Estimation, and Compensation

Springer

Yabo Li
Department of Information Science
and Electronic Engineering
Zhejiang University
Hangzhou, Zhejiang
P.R. China

ISSN 2191-8112 ISSN 2191-8120 (electronic)
ISBN 978-1-4614-8617-6 ISBN 978-1-4614-8618-3 (eBook)
DOI 10.1007/978-1-4614-8618-3
Springer New York Heidelberg Dordrecht London

Library of Congress Control Number: 2013949282

Printed on acid-free paper

Springer is part of Springer Science+Business Media (www.springer.com)

Preface

In wireless communication systems, In-phase and Quadrature (IQ) modulator and demodulator are usually used at transmitter (TX) and receiver (RX), respectively. For Digital-to-Analog Converter (DAC) and Analog-to-Digital Converter (ADC) limited systems, such as multi-giga-hertz bandwidth millimeter-wave systems, using analog modulator and demodulator is still a low power and low cost solution. In this kind of systems, the IQ imbalance cannot be ignored. Numerous papers have been published to investigate this problem. However, depending on different assumptions, different models of IQ imbalance are built and different approaches are proposed. This makes the researchers and system design engineers who are new to this topic difficult to start with. The goal of this book is to provide a unified IQ imbalance model and systematically review the existing estimation and compensation schemes. The intended audience of this book is the researchers who are interested in the IQ imbalance as well as the system design engineers who need to deal with the IQ imbalance in their systems.

The book starts with a unified IQ imbalance model, and then investigates the methods that are used in estimation and compensation. Since different methods may depend on different system assumptions, application scenarios, and implementation architectures, it is difficult to compare the performance in a fair setup. So, when discussing these methods, the book mainly focuses on the mathematical derivations. The hope is that after reading this book the readers can find an existing method or devise a new one that is more suitable for the systems they are going to design. Due to the limited space and time, and also the limited knowledge of the author, the method discussed in this book is far from extensive, and only literatures that the author is familiar with are included. The author would like to apologize if any important literatures are missing in the references.

The author would like to thank Dr. Xuemin (Sherman) Shen for providing this opportunity to write this book.

Hangzhou, P.R. China
Yabo Li

Acknowledgements

The work is supported by the National Science Foundation of China (NSFC) under Grant 61010197 and 61271243, Chinese Ministry of Education New Faculty Fund under Grant 20100101120039, and National Science and Technology Major Project under Grant 2011ZX03003-003-03.

Contents

Acronyms

ADC	Analog-to-Digital Converter
BLAST	Bell-Labs Layered Space-Time
CP	Cyclic Prefix
CRE	Channel Residue Energy
DAC	Digital-to-Analog Converter
DDC	Digital Down Converter
DFT	Discrete Fourier Transform
EM	Expectation Maximization
EVM	Error Vector Magnitude
FFT	Fast Fourier Transform
FI	Frequency Independent
GI	Guard Interval
GPP	Generalized Periodic Preamble
ICI	Inter-Carrier Interference
IDFT	Inverse Discrete Fourier Transform
IF	Intermediate Frequency
IFFT	Inverse Fast Fourier Transform
IQ	In-Phase and Quadrature
LMMSE	Linear Minimum Mean Square Error
LMS	Least Mean Square
LS	Least Square
LPF	Low Pass Filter
LTE	Long Term Evolution
MIMO	Multiple-Input Multiple-Output
MPP	Modified Periodic Preamble
OFDM	Orthogonal-Frequency Division Multiplexing
PSK	Phase-Shift Keying
QAM	Quadrature Amplitude Modulation
RF	Radio Frequency
RLS	Recursive Least Square
RX	Receiver

SC-FDE	Single Carrier Frequency Domain Equalization
SFC	Space-Frequency Code
SIR	Signal-to-Interference Ratio
STC	Space-Time Code
TX	Transmitter
WLAN	Wireless Local Area Network

Chapter 1
Introduction

Abstract In this chapter, we introduce what is in-phase and quadrature (IQ) imbalance, where it comes from, and what kinds of aspects of IQ imbalance have been discussed in the literature.

In wireless communication systems, it is inevitable to use analog components, such as filters, mixers, and amplifiers, etc. These analog components cannot be made ideal, due to the limitations in costs, power consumptions, or chip areas. The non-ideality of the analog components impairs the system performance, and most of them cannot be removed in the analog domain. So, designing schemes to compensate the impairments in the digital domain is necessary. When the carrier frequency is ultra-high and/or the bandwidth is ultra-wide, the analog impairments may dominate the system performance. For example, in IEEE 802.11ad systems [1], multi-giga-hertz bandwidth in 60 GHz band is used, when designing such systems, to achieve high performance and low cost, digital compensation is a must [2].

In-phase and Quadrature (IQ) imbalance, caused by using analog IQ modulator and/or demodulator, destroys the orthogonality between the signals in I and Q branches, and degrades the signal-to-distortion ratio of the received signal [3]. When carrier frequency is low and bandwidth is narrow, directly digitizing the RF (Radio Frequency) or IF (Intermediate Frequency) signal is viable. In this case, if digital IQ modulation and demodulation are used, IQ imbalance does not exist. However, for the systems that are digital-to-analog converter (DAC) or analog-to-digital converter (ADC) limited, such as IEEE 802.11ad, directly digitizing the RF or IF signal is very costly, if not impossible. In this case, using analog IQ modulator and demodulator is still a cheap solution. So, investigating methods to efficiently estimate and compensate the IQ imbalance is still necessary.

In wireless communication systems, both non-ideal analog IQ modulator and demodulator may introduce IQ imbalance, so IQ imbalance exists in both transmitter (TX) and receiver (RX). Also, in either TX or RX, IQ imbalance may come from

Y. Li, *In-Phase and Quadrature Imbalance: Modeling, Estimation, and Compensation*, SpringerBriefs in Electrical and Computer Engineering, DOI 10.1007/978-1-4614-8618-3_1,

two different sources, one is from the mixer, which is frequency independent (FI), the other is from the mismatched analog filters in the I and Q branches, which is frequency dependent (FD).

The IQ imbalance has been extensively discussed in the past several years, please see the references in the following chapters. Depending on different system assumptions, different aspects of the IQ imbalance are discussed. If the system is narrow band, where the frequency dependence of IQ imbalance is negligible, there are papers that only considered FI IQ imbalance. If the system is wideband, where the IQ imbalance is frequency selective, there are papers that considered FD IQ imbalance. If only transmitter is designed to meet the error vector magnitude (EVM) specification, there are papers that considered TX only IQ imbalance. If only RX is designed, which is inter-operable with other standard compliant transmitters, there are papers that considered RX only IQ imbalance. If TX and RX are jointly designed, there are papers that investigated joint TX and RX IQ imbalances. Also, depending on different transmission schemes, there are papers that discussed the IQ imbalance for single carrier systems, and there are also papers that discussed the IQ imbalance for OFDM (Orthogonal Frequency Division Multiplexing) systems. In a system, most often, not only IQ imbalance is present, so there are also papers that discussed IQ imbalance together with frequency offset and/or phase noise. Since multiple antennas are more and more popular in wireless communication systems, there are papers that investigated the IQ imbalance for multiple antenna systems. In most cases, the estimation and compensation of IQ imbalance rely on a specially designed preamble, however, there are papers that investigated blind IQ imbalance estimation and compensation.

For all these aspects of IQ imbalance, different papers may start from different IQ imbalance model, and develop their approaches based on different assumptions. In this short book, a unified IQ imbalance model is built first, then different methods are reviewed and derived based on this unified model. It gives a clear picture about how different methods are developed, in what conditions they apply, and how different methods proposed in different papers are related. This helps the researchers who are interested in but are new to this topic to start their research quickly, it also helps the system designers who are trying to deal with the IQ imbalance in their systems to find an appropriate existing method or devise a new method.

The book is organized as follows. In Chap. 2, a unified model is built. Different approaches to deal with FI IQ imbalance are discussed in Chap. 3. Methods to estimate and compensate FD IQ imbalance are discussed in Chap. 4.

References

1. "IEEE 802.11ad standard draft D0.1," [Available] http://www.ieee802.org/11/Reports/tgadupdate.htm.
2. M. Dohler, R. W. Heath, A. Lozano, C. B. Papadias, and R. A. Valenzuela, "Is the PHY layer dead?," *IEEE Commun. Mag.*, pp. 159–165, Apr. 2011.
3. B. Razavi, *RF Microelectronics*, Upper Saddle River, NJ: Prentice Hall, 1998.

Chapter 2
The IQ Imbalance Model

Abstract Both analog In-phase and Quadrature (IQ) modulator and demodulator may introduce IQ imbalance. If it is left uncompensated, the system performance will be impaired. In this chapter, we first derive the IQ imbalance model without any other impairments. Then, we consider the IQ imbalance model in presence of frequency offset and phase noise. After that, we consider the IQ imbalance model in multiple antenna systems, firstly space-time or space-frequency encoded systems, then spatial multiplexing systems. At the end, we show the signal-to-interference ratio (SIR) degradation due to the IQ imbalance.

2.1 Introduction

In-phase and quadrature (IQ) modulator and demodulator are commonly used in wireless communication systems [1]. The IQ modulator at transmitter transforms the complex baseband signals to passband centered at carrier frequency, which is suitable for wireless transmission, while the IQ demodulator transforms the passband signals to complex baseband, which is suitable for using signal processing techniques to recover the transmitted signals. Ideal IQ modulator and demodulator provide two orthogonal channels for the real and imaginary parts of a complex signal. In reality, the IQ modulator and demodulator are not ideal, especially for wireless transceiver designed with direct conversion radio frequency (RF) architectures [2]. This introduces interference between the two orthogonal channels. If it is left uncompensated, the system may not have sufficient performance to support high order modulation schemes, and thus may not be able to support high data rate. To compensate IQ imbalance, we first need to build an IQ imbalance model. In this chapter, we start with general frequency dependent (FD) IQ imbalance model without other impairments. Then we consider the the IQ imbalance model when there are frequency offset and phase noise. Considering multiple antennas, we show the IQ imbalance model for space-time/space-frequency encoded systems

Y. Li, *In-Phase and Quadrature Imbalance: Modeling, Estimation, and Compensation*, SpringerBriefs in Electrical and Computer Engineering, DOI 10.1007/978-1-4614-8618-3_2,

as well as the IQ imbalance model for spatial multiplexing systems. This chapter prepares for the discussion of estimation and compensation in the next two chapters.

2.2 The IQ Imbalance Model Without Other Impairments

In this section, we derive the IQ imbalance model without other impairments. This IQ imbalance model can be found in almost all literatures about IQ imbalance estimation and compensation, see for example [3–15] and the references therein. We unify these models in this chapter. We start from general FD time domain IQ imbalance model, then the frequency independent (FI) one is just a special case. We then look at the FD IQ model in the frequency domain assuming that OFDM is used.

2.2.1 The Time Domain Transmitter-Side IQ Imbalance Model

Figure 2.1 shows the IQ modulation with amplitude and phase imbalances. In the system, there are two sources that may cause IQ imbalance. One is the phase and amplitude mismatch in the clocks used by the I and Q branches. This mismatch is constant for different frequency components in the transmitted signal, and is called FI IQ imbalance. The other is caused by the discrepancy in the filters used by the I and Q branches. This mismatch is different for different frequency components in the transmitted signal, and is called FD IQ imbalance.

At the transmitter, we assume that the clocks used by the I and Q branches have amplitude mismatch ε_t and phase mismatch θ_t, the discrete time impulse response of the analog filters in the I and Q branches are $g_t^I(n)$ and $g_t^Q(n)$, respectively, and the discrete time signals to be transmitted in the I and Q branches are $s^I(n)$ and $s^Q(n)$, respectively. Then, the modulated passband signal $x_p(n)$ equals

$$\begin{aligned} x_p(n) &= \left[s^I(n) \otimes g_t^I(n)\right](1-\varepsilon_t)\cos(\omega_c n - \theta_t) \\ &\quad - \left[s^Q(n) \otimes g_t^Q(n)\right](1+\varepsilon_t)\sin(\omega_c n + \theta_t) \\ &= x^I(n)\cos(\omega_c n) - x^Q(n)\sin(\omega_c n), \end{aligned}$$

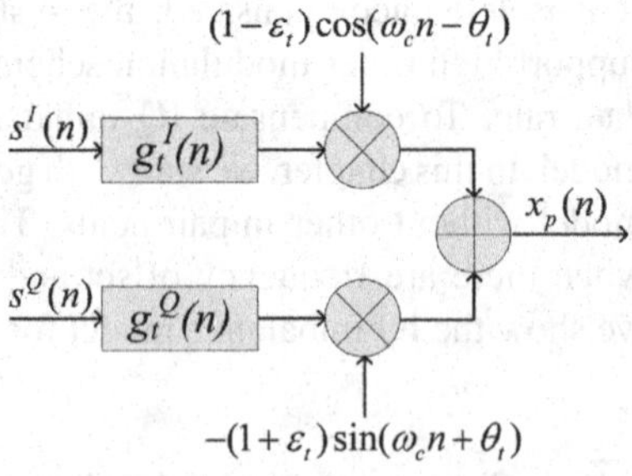

Fig. 2.1 The transmitter IQ imbalance signal model

where ω_c is the carrier frequency, $\otimes$ means convolution, and $x^I(n)$ and $x^Q(n)$ are defined respectively as

$$x^I(n) \triangleq \left[s^I(n) \otimes g_t^I(n)\right](1-\varepsilon_t)\cos\theta_t - \left[s^Q(n) \otimes g_t^Q(n)\right](1+\varepsilon_t)\sin\theta_t, \tag{2.1}$$

$$x^Q(n) \triangleq -\left[s^I(n) \otimes g_t^I(n)\right](1-\varepsilon_t)\sin\theta_t + \left[s^Q(n) \otimes g_t^Q(n)\right](1+\varepsilon_t)\cos\theta_t. \tag{2.2}$$

Writing the complex baseband signal to be transmitted as

$$s(n) \triangleq s^I(n) + \mathbf{j}\, s^Q(n),$$

and the TX IQ imbalance distorted complex baseband signal as

$$x'(n) \triangleq x^I(n) + \mathbf{j}\, x^Q(n),$$

based on (2.1) and (2.2), we have

$$x'(n) = \mu_t(n) \otimes s(n) + \nu_t(n) \otimes s^*(n), \tag{2.3}$$

where

$$\mu_t(n) \triangleq \left(\frac{\alpha_t - \beta_t}{2}\right) g_t^I(n) + \left(\frac{\alpha_t + \beta_t}{2}\right) g_t^Q(n), \tag{2.4}$$

$$\nu_t(n) \triangleq \left(\frac{\alpha_t - \beta_t}{2}\right) g_t^I(n) - \left(\frac{\alpha_t + \beta_t}{2}\right) g_t^Q(n), \tag{2.5}$$

and α_t and β_t are defined as

$$\alpha_t \triangleq \cos\theta_t + \mathbf{j}\,\varepsilon_t \sin\theta_t, \tag{2.6}$$

$$\beta_t \triangleq \varepsilon_t \cos\theta_t + \mathbf{j}\,\sin\theta_t, \tag{2.7}$$

respectively.

If there is no FD IQ imbalance, i.e., assume that $g_t(n) = g_t^I(n) = g_t^Q(n)$, then we have

$$\mu_t(n) = \alpha_t g_t(n)$$
$$\nu_t(n) = -\beta_t g_t(n).$$

Substituting them into (2.3), we have

$$x'(n) = \left[\alpha_t s(n) - \beta_t s^*(n)\right] \otimes g_t(n). \tag{2.8}$$

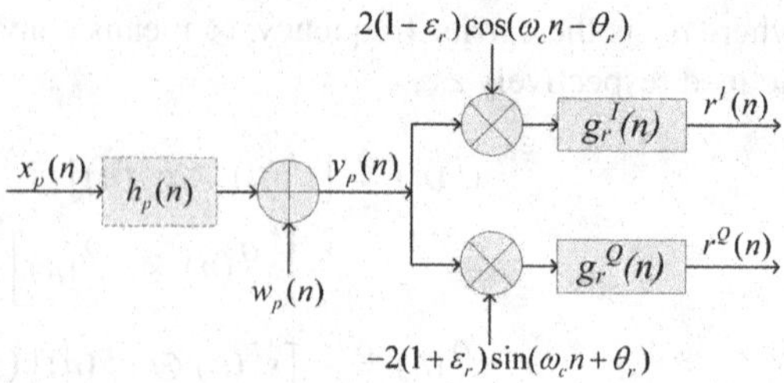

Fig. 2.2 The receiver IQ imbalance signal model

2.2.2 The Time Domain Receiver-Side IQ Imbalance Model

Figure 2.2 shows the IQ demodulation with amplitude and phase imbalances. Also, in the system, we assume that both FI IQ imbalance caused by mixer and FD IQ imbalance caused by filters are present.

At the receiver, the received passband signal can be written as

$$y_p(n) = x_p(n) \otimes h_p(n) + w_p(n), \tag{2.9}$$

where $h_p(n)$ is the discrete time passband channel response and $w_p(n)$ is the passband additive noise. If we denote $y^I(n)$ and $y^Q(n)$ as the output of ideal IQ demodulator at the I and Q branches, respectively, then, $y_p(n)$ can be written as

$$y_p(n) = y^I(n)\cos(\omega_c n) - y^Q(n)\sin(\omega_c n), \tag{2.10}$$

and its equivalent complex baseband signal can be written as

$$\begin{aligned} y'(n) &\triangleq y^I(n) + \mathbf{j}\, y^Q(n) \\ &= x'(n) \otimes h'(n) + w'(n), \end{aligned} \tag{2.11}$$

where $h'(n)$ is the equivalent complex baseband channel response, and $w'(n)$ is the equivalent complex noise with two-sided power spectrum density σ^2.

Based on the RX IQ demodulation model in Fig. 2.2, with amplitude mismatch ε_r and phase mismatch θ_r at RX mixer and analog filters with discrete time impulse response $g_r^I(n)$ and $g_r^Q(n)$ at I and Q branches, respectively, we have that the demodulated output at I and Q branches equals to

$$r^I(n) = \text{LPF}\left\{\left[2y_p(n)(1-\varepsilon_r)\cos(\omega_c n - \theta_r)\right] \otimes g_r^I(n)\right\}, \tag{2.12}$$

$$r^Q(n) = \text{LPF}\left\{\left[-2y_p(n)(1+\varepsilon_r)\sin(\omega_c n + \theta_r)\right] \otimes g_r^Q(n)\right\}, \tag{2.13}$$

respectively, where the operation LPF$\{\cdot\}$ removes the frequency content at $2\omega_c$. Substituting (2.10) into (2.12) and (2.13), expanding (2.12) and (2.13), and removing the high frequency signals at $2\omega_c$, we have

$$r^I(n) = \left[y^I(n) \otimes g_r^I(n)\right](1-\varepsilon_r)\cos\theta_r$$
$$-\left[y^Q(n) \otimes g_r^I(n)\right](1-\varepsilon_r)\sin\theta_r \quad (2.14)$$
$$r^Q(n) = -\left[y^I(n) \otimes g_r^Q(n)\right](1+\varepsilon_r)\sin\theta_r$$
$$+\left[y^Q(t) \otimes g_r^Q(n)\right](1+\varepsilon_r)\cos\theta_r. \quad (2.15)$$

Writing the RX IQ imbalance distorted complex baseband signal as

$$r(n) \triangleq r^I(n) + \mathbf{j}\, r^Q(n),$$

we have that $r(n)$ equals to

$$r(n) = \mu_r(n) \otimes y'(n) + \nu_r(n) \otimes (y'(n))^*, \quad (2.16)$$

where $\mu_r(n)$ and $\nu_r(n)$ are

$$\mu_r(n) \triangleq \left(\frac{\alpha_r - \beta_r^*}{2}\right) g_r^I(n) + \left(\frac{\alpha_r + \beta_r^*}{2}\right) g_r^Q(n), \quad (2.17)$$
$$\nu_r(n) \triangleq \left(\frac{\alpha_r^* - \beta_r}{2}\right) g_r^I(n) - \left(\frac{\alpha_r^* + \beta_r}{2}\right) g_r^Q(n), \quad (2.18)$$

and α_r and β_r are defined as

$$\alpha_r \triangleq \cos\theta_r - \mathbf{j}\,\varepsilon_r \sin\theta_r, \quad (2.19)$$
$$\beta_r \triangleq \varepsilon_r \cos\theta_r + \mathbf{j}\, \sin\theta_r. \quad (2.20)$$

If there is no FD IQ imbalance, i.e., assume $g_r(n) = g_r^I(n) = g_r^Q(n)$, we have

$$\mu_r(n) = \alpha_r g_r(n)$$
$$\nu_r(n) = -\beta_r g_r(n).$$

Substitute them into (2.16) gives

$$r(n) = \left[\alpha_r y'(n) - \beta_r (y'(n))^*\right] \otimes g_r(n). \quad (2.21)$$

2.2.3 *The Time Domain Transmitter and Receiver IQ Imbalance Model*

Now, we assume that in the system, IQ imbalance exists in both transmitter and receiver. Substituting (2.3) into (2.11), we have

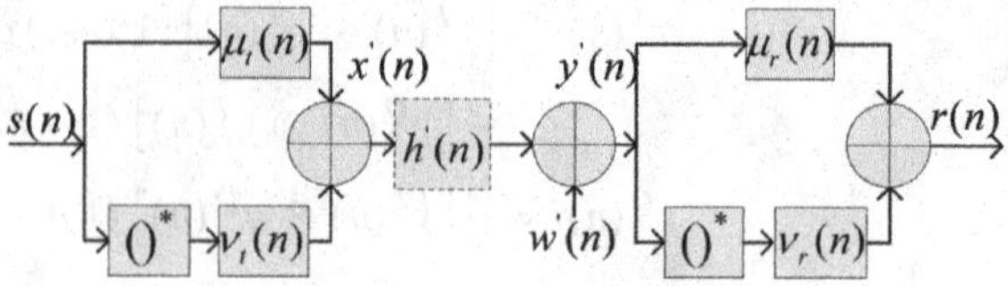

Fig. 2.3 Transmitter and receiver IQ imbalance complex signal model

$$y'(n) = [\mu_t(n) \otimes s(n) + \nu_t(n) \otimes s^*(n)] \otimes h'(n) + w'(n), \tag{2.22}$$

and further substituting (2.22) into (2.16), we have

$$\begin{aligned} r(n) = &\left[\mu_r(n) \otimes \mu_t(n) \otimes h'(n) + \nu_r(t) \otimes \nu_t^*(n) \otimes (h'(n))^*\right] \otimes s(n) \\ &+ \left[\mu_r(n) \otimes \nu_t(n) \otimes h'(n) + \nu_t(t) \otimes \mu_t^*(n) \otimes (h'(n))^*\right] \otimes s^*(n) \\ &+ \left[\mu_r(n) \otimes w'(n) + \nu_r(n) \otimes (w'(n))^*\right]. \end{aligned} \tag{2.23}$$

Figure 2.3 shows the time domain model which includes both transmitter and receiver IQ imbalances.

To simplify the signal model, we can absorb the filtering operations that are common to both I and Q branches in the transmitter and receiver to the channel. This will make the design of the estimation and compensation schemes easier. To do so, we can write $y'(n)$ in (2.22) as

$$\begin{aligned} y'(n) &= [\mu_t(n) \otimes s(n) + \nu_t(n) \otimes s^*(n)] \otimes h'(n) + w'(n) \\ &= [s(n) + \underbrace{\nu_t(n) \otimes \overline{\mu}_t(n)}_{\triangleq \xi_t(n)} \otimes s^*(n)] \otimes \underbrace{\mu_t(n) \otimes h'(n)}_{\triangleq h''(n)} + w'(n), \end{aligned}$$

where $\overline{\mu}_t(n)$ satisfies

$$\overline{\mu}_t(n) \otimes \mu_t(n) = \delta(t)$$

and $\delta(n)$ is the delta function, which equals

$$\delta(n) = \begin{cases} 1, & n = 0 \\ 0, & n \neq 0 \end{cases}.$$

If we denote

$$x(n) \triangleq s(n) + \xi_t(n) \otimes s^*(n), \tag{2.24}$$

we can write $y'(n)$ as

$$y'(n) = x(n) \otimes h''(n) + w'(n). \tag{2.25}$$

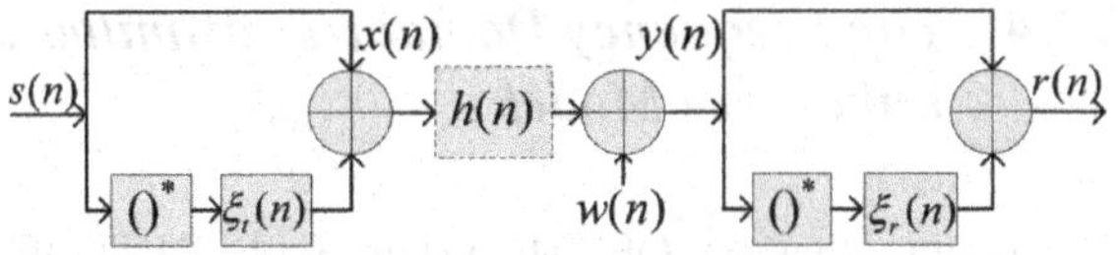

Fig. 2.4 IQ imbalance equivalent complex signal model

If we define

$$y(n) \triangleq \mu_r(n) \otimes y'(n), \tag{2.26}$$

and substitute (2.25) into (2.26), we have

$$\begin{aligned} y(n) &= \mu_r(n) \otimes (h''(n) \otimes x(n) + w'(n)) \\ &= \underbrace{\mu_r(n) \otimes h'(n) \otimes \mu_t(n)}_{\triangleq h(n)} \otimes x(n) + \underbrace{\mu_r(n) \otimes w'(n)}_{\triangleq w(n)} \\ &= h(n) \otimes x(n) + w(n). \end{aligned} \tag{2.27}$$

Then, the received signal $r(n)$ in Eq. (2.16) becomes

$$r(n) = y(n) + \underbrace{\nu_r(n) \otimes \overline{\mu}_r^*(n)}_{\triangleq \xi_r(n)} \otimes y^*(n) \tag{2.28}$$

$$= y(n) + \xi_r(n) \otimes y^*(n), \tag{2.29}$$

where $\overline{\mu}_r^*(n)$ satisfies that

$$\overline{\mu}_r^*(n) \otimes \mu_r^*(n) = \delta(n).$$

Figure 2.4 shows the equivalent complex signal model from the transmitted signal $s(n)$ to the received signal $r(n)$, where $x(n)$ is related to $s(n)$ based on (2.24), $y(n)$ is related to $x(n)$ based on (2.27), and $r(n)$ is related to $y(n)$ based on (2.29).

Substitute (2.24) into (2.27), and then substitute (2.27) into (2.29), we can get the relation between the transmit signal $s(n)$ and the received signal $r(n)$, which is

$$\begin{aligned} r(n) = {} & h(n) \otimes s(n) + h(n) \otimes \xi_t(n) \otimes s^*(n) \\ & + \xi_r(n) \otimes h^*(n) \otimes s^*(n) + \xi_r(n) \otimes h^*(n) \otimes \xi_t^*(n) \otimes s(n) \\ & + w(n) + \xi_r(n) \otimes w^*(n). \end{aligned} \tag{2.30}$$

2.2.4 The Frequency Domain Transmitter and Receiver IQ Imbalance Model

We assume that an OFDM system with DFT/IDFT (Discrete Fourier Transform/Inverse Discrete Fourier Transform) size of N is used, where N is usually a power of 2 and thus an even number. Assume that $\xi_t(n)$, $h(n)$ and $\xi_r(n)$ have length L_t, L_h and L_r, respectively. Assume that the cyclic prefix length $L_{cp} > L_t + L_h + L_r - 2$. After CP removal, define the N sampled value of $x(n)$ and $s(n)$ in one OFDM symbol respectively as

$$\bar{\mathbf{x}} = [x(0), x(1), \cdots, x(N-1)]^T,$$

$$\bar{\mathbf{s}} = [s(0), s(1), \cdots, s(N-1)]^T.$$

Then, based on (2.24), we have

$$\bar{\mathbf{x}} = \bar{\mathbf{s}} + \Lambda_t \bar{\mathbf{s}}^*, \tag{2.31}$$

where, $\bar{\mathbf{s}}^*$ is the component-wise conjugate of $\mathbf{s}$, and Λ_t is an $N \times N$ circulant matrix with the first column equals $\left[\xi_t(0), \xi_t(1), \cdots, \xi_t(L_t - 1), \mathbf{0}_{1\times(N-L_t)}\right]^T$, where $\mathbf{0}_{1\times(N-L_t)}$ is a size $N - L_t$ row vector with all zero elements.

Decompose Λ_t as

$$\Lambda_t = \mathbf{F}^H \Xi_t \mathbf{F}, \tag{2.32}$$

where

$$\Xi_t \triangleq \operatorname{diag}\{\xi_{t,0}, \xi_{t,1}, \cdots, \xi_{t,N-1}\},$$

$\xi_{t,k} \triangleq \sum_{n=0}^{N-1} \xi_t(n) e^{-j\frac{2\pi nk}{N}}$, and $\mathbf{F}$ is an $N \times N$ DFT matrix.

Substitute (2.32) into (2.31), and multiply both sizes of (2.31) by $\mathbf{F}$, we have

$$\mathbf{x} = \mathbf{s} + \Xi_t \mathbf{s}^\#, \tag{2.33}$$

where $\mathbf{x}$, $\mathbf{s}$ and $\mathbf{s}^\#$ are defined respectively as

$$\mathbf{x} = [X_0, X_1, \cdots, X_{N-1}]^T,$$
$$\mathbf{s} = [S_0, S_1, \cdots, S_{N-1}]^T,$$
$$\mathbf{s}^\# = [S_0^*, S_{N-1}^*, S_{N-2}^*, \cdots, S_1^*]^T,$$

and X_k and S_k are defined respectively as $X_k = \sum_{n=0}^{N-1} x(n)e^{-j\frac{2\pi nk}{N}}$ and $S_k = \sum_{n=0}^{N-1} s(n)e^{-j\frac{2\pi nk}{N}}$. Expanding (2.33) gives us

$$X_k = S_k + \xi_{t,k} S_{N-k}^*,$$
$$X_{N-k} = S_{N-k} + \xi_{t,N-k} S_k^*,$$

for $k = 1, 2, \cdots, N-1$ or in matrix format

$$\begin{bmatrix} X_k \\ X_{N-k}^* \end{bmatrix} = \begin{bmatrix} 1 & \xi_{t,k} \\ \xi_{t,N-k}^* & 1 \end{bmatrix} \begin{bmatrix} S_k \\ S_{N-k}^* \end{bmatrix}. \tag{2.34}$$

Similarly, based on (2.27), after removing CP, and multiplying size $N \times N$ DFT matrix $\mathbf{F}$ on both sides of (2.27), we have

$$\begin{bmatrix} Y_k \\ Y_{N-k}^* \end{bmatrix} = \begin{bmatrix} H_k & 0 \\ 0 & H_{N-k}^* \end{bmatrix} \begin{bmatrix} X_k \\ X_{N-k}^* \end{bmatrix} + \begin{bmatrix} W_k \\ W_{N-k}^* \end{bmatrix}, \tag{2.35}$$

where $Y_k = \sum_{n=0}^{N-1} y(n)e^{-j\frac{2\pi nk}{N}}$, $H_k = \sum_{n=0}^{L_h-1} h(n)e^{-j\frac{2\pi nk}{N}}$, and $W_k = \sum_{n=0}^{N-1} w(n)e^{-j\frac{2\pi nk}{N}}$. Based on (2.27), W_k can be further written as

$$W_k = \mu_{r,k} W_k', \tag{2.36}$$

where $\mu_{r,k}$ and W_k equal to $\mu_{r,k} = \sum_{n=0}^{N-1} \mu_r(n)e^{-j\frac{2\pi nk}{N}}$ and $W_k' = \sum_{n=0}^{N-1} w'(n)e^{-j\frac{2\pi kn}{N}}$, respectively.

Also, following the same procedure, based on (2.29), we have

$$\begin{bmatrix} R_k \\ R_{N-k}^* \end{bmatrix} = \begin{bmatrix} 1 & \xi_{r,k} \\ \xi_{r,N-k}^* & 1 \end{bmatrix} \begin{bmatrix} Y_k \\ Y_{N-k}^* \end{bmatrix} \tag{2.37}$$

$$= \mathbf{H}_k \begin{bmatrix} S_k \\ S_{N-k}^* \end{bmatrix} + \mathbf{v}_k, \tag{2.38}$$

where $R_k = \sum_{n=0}^{N-1} r(n)e^{-j\frac{2\pi kn}{N}}$ and $\xi_{r,k} = \sum_{n=0}^{L_r-1} \xi_r(n)e^{-j\frac{2\pi kn}{N}}$. $\mathbf{H}_k$ is the equivalent channel matrix that combines channel response, transmitter and receive IQ imbalances for subcarrier pair k and $N-k$, and can be written as

$$\mathbf{H}_k = \begin{bmatrix} 1 & \xi_{r,k} \\ \xi_{r,N-k}^* & 1 \end{bmatrix} \begin{bmatrix} H_k & 0 \\ 0 & H_{N-k}^* \end{bmatrix} \begin{bmatrix} 1 & \xi_{t,k} \\ \xi_{t,N-k}^* & 1 \end{bmatrix}$$
$$= \begin{bmatrix} H_k + \xi_{r,k}\xi_{t,N-k}^* H_{N-k}^* & \xi_{t,k} H_k + \xi_{r,k} H_{N-k}^* \\ \xi_{r,N-k}^* H_k + \xi_{t,N-k}^* H_{N-k}^* & \xi_{r,N-k}^* \xi_{t,k} H_k + H_{N-k}^* \end{bmatrix}. \tag{2.39}$$

$\mathbf{v}_k$ is the colored noise defined as

$$\begin{aligned}
\mathbf{v}_k &= \begin{bmatrix} V_k \\ V_{N-k}^* \end{bmatrix} = \begin{bmatrix} 1 & \xi_{r,k} \\ \xi_{r,N-k}^* & 1 \end{bmatrix} \begin{bmatrix} W_k \\ W_{N-k}^* \end{bmatrix} \\
&= \begin{bmatrix} 1 & \xi_{r,k} \\ \xi_{r,N-k}^* & 1 \end{bmatrix} \begin{bmatrix} \mu_{r,N-k} & 0 \\ 0 & \mu_{r,k}^* \end{bmatrix} \begin{bmatrix} W_k' \\ (W_{N-k}')^* \end{bmatrix} \\
&= \begin{bmatrix} \mu_{r,N-k} & \nu_{r,k} \\ \nu_{r,N-k}^* & \mu_{r,k}^* \end{bmatrix} \begin{bmatrix} W_k' \\ (W_{N-k}')^* \end{bmatrix},
\end{aligned} \tag{2.40}$$

where $\nu_{r,k} = \sum_{n=0}^{N-1} \nu_r(n) e^{-j\frac{2\pi kn}{N}}$. In (2.40), we used the definition $\xi_r(n) \triangleq \nu_r(n) \otimes \overline{\mu}_r^*(n)$ and the fact that $\xi_{r,k} = \nu_{r,k}/\mu_{r,k}^*$. From (2.40) we can see that the covariance of the noise depends on the individual values of $\nu_{r,k}$ and $\mu_{r,k}$, not just their ratio.

2.3 The IQ Imbalance Model with Frequency Offset

In practical wireless communication system, the frequency of local oscillator at the transmitter and receiver may not be the same, this introduces frequency offset. In presence of frequency offset, the IQ imbalance model is different from that derived in Sect. 2.2. There are also many literatures that discussed IQ imbalance with frequency offset, see for example [15–22] and the references therein. In this section, we first look at the IQ imbalance model in presence of frequency offset in the time domain, and then look at it in the frequency domain.

Assume that the carrier frequency of the receiver is offset by Δf with respect to the carrier frequency of the transmitter, let us denote $\Delta\omega \triangleq 2\pi \Delta f T$, where T is the sample period. Then, at the receiver, the frequency offset causes the baseband equivalent signal before the IQ demodulation rotated by an angle $e^{j\Delta\omega n}$, i.e., instead of $y'(n)$ defined in (2.11) is received, $e^{j\Delta\omega n} y'(n)$ is received. After IQ demodulation, based on (2.16), the baseband equivalent IQ imbalance distorted signal $r(n) = r^I(n) + \mathbf{j} r^Q(n)$ equals

$$\begin{aligned}
r(n) &= \mu_r(n) \otimes \left[e^{j\Delta\omega n} y'(n)\right] + \nu_r(n) \otimes \left[e^{j\Delta\omega n} y'(n)\right]^* \\
&= e^{j\Delta\omega n} \left[\left(e^{-j\Delta\omega n} \mu_r(n)\right) \otimes y'(n)\right] \\
&\quad + e^{-j\Delta\omega n} \left[\left(e^{j\Delta\omega n} \nu_r(n)\right) \otimes (y'(n))^*\right],
\end{aligned} \tag{2.41}$$

where $\mu_r(n)$ and $\nu_r(n)$ are defined as Eqs. (2.19) and (2.20), respectively.

From (2.41) we can see that when RX IQ exists, we cannot multiply $e^{-j\Delta\omega n}$ to $r(n)$ to compensate the frequency offset, since there are $e^{-j\Delta\omega n}$ component in $r(n)$, which cannot be removed by multiplying $e^{-j\Delta\omega n}$. So, at the receiver, before compensating frequency offset, the RX IQ imbalance must be compensated.

When both transmitter and receiver IQ imbalances exist, the baseband equivalent received signal before IQ demodulation equals (2.22), which is

$$y'(n) = [\mu_t(n) \otimes s(n) + \nu_t(n) \otimes s^*(n)] \otimes h'(n) + w'(n). \tag{2.42}$$

Substituting it into (2.41) gives

$$\begin{aligned} r(n) = e^{j\Delta\omega n}\{&\mu_t(n) \otimes h'(n) \otimes [e^{-j\Delta\omega n}\mu_r(n)] \otimes s(n)\} \\ &+ e^{j\Delta\omega n}\{\nu_t(n) \otimes h'(n) \otimes [e^{-j\Delta\omega n}\mu_r(n)] \otimes s^*(n)\} \\ &+ e^{-j\Delta\omega n}\{\mu_t^*(n) \otimes (h'(n))^* \otimes [e^{j\Delta\omega n}\nu_r(n)] \otimes s^*(n)\} \\ &+ e^{-j\Delta\omega n}\{\nu_t^*(n) \otimes (h'(n))^* \otimes [e^{j\Delta\omega n}\nu_r(n)] \otimes s(n)\} \\ &+ q(n), \end{aligned} \tag{2.43}$$

where $q(n)$ is the transformed noise, which equals

$$q(n) = \mu_r(n) \otimes \left[e^{j\Delta\omega n}w'(n)\right] + \nu_r(n) \otimes \left[e^{j\Delta\omega n}w'(n)\right]^*. \tag{2.44}$$

From (2.43) we can see how the frequency offset is added to the received signal. If there is no frequency offset, i.e., $\Delta\omega = 0$, then (2.43) is equivalent to (2.23). If there is no transmitter and receiver IQ imbalance, we have $\mu_t(n) = \mu_r(n) = \delta(n)$ and $\nu_t(n) = \nu_r(n) = 0$. Substituting them into (2.43), we have $r(n) = e^{j\Delta\omega n}[h'(n) \otimes s(n) + w'(n)]$, which is a standard received signal model with frequency offset.

Also, to simplify the notations, we can absorb the filtering operations that are common to the I and Q branches to the channel. In this case, the received signal before IQ demodulator can still be represented as

$$y'(n) = x(n) \otimes h''(n) + w'(n), \tag{2.45}$$

where

$$\begin{aligned} x(n) &= s(n) + \xi_t(n) \otimes s^*(n) \\ h''(n) &= \mu_t(n) \otimes h'(n) \end{aligned} \tag{2.46}$$

the same as that in Sect. 2.2.3. If we define

$$y(n) = \mu_r(n) \otimes \left[e^{j\Delta\omega n}y'(n)\right], \tag{2.47}$$

then $r(n)$ in (2.41) can be represented as

$$r(n) = y(n) + \xi_r(n) \otimes y^*(n), \tag{2.48}$$

which is the same as Eq. (2.29) in Sect. 2.2.3.

Compared with the IQ model without frequency offset, the difference here is in $y(n)$, which is different from that in Eq. (2.26) in Sect. 2.2.3. Substituting $y'(n) = x(n) \otimes h''(n) + w'(n)$ into the new $y(n)$ in (2.47), we have

$$\begin{aligned} y(n) &= e^{j\Delta\omega n}\left[\left(e^{-j\Delta\omega n}\mu_r(n)\right) \otimes y'(n)\right] \\ &= e^{j\Delta\omega n}\left\{x(n) \otimes \underbrace{\mu_t(n) \otimes h'(n) \otimes [e^{-j\Delta\omega n}\mu_r(n)]}_{\triangleq h(n)} + \underbrace{w'(n) \otimes [e^{-j\Delta\omega n}\mu_r(n)]}_{\triangleq w(n)}\right\} \\ &= e^{j\Delta\omega n}\{x(n) \otimes h(n) + w(n)\}. \end{aligned} \tag{2.49}$$

Compared with the $y(n)$ defined in (2.26), we can find that the frequency offset affects the equivalent channel $h(n)$ as well as the equivalent noise $w(n)$.

Now, we consider the IQ imbalance in the frequency domain with frequency offset. We assume that the channel impulse response keeps constant during one OFDM symbol period, and look at the impact of frequency offset on the IQ imbalance model. We assume that size N FFT/IFFT is used and the CP is greater than the length of the equivalent channel. After removing CP, we write the $y(n)$ in one OFDM symbol as

$$\overline{\mathbf{y}} = [y(0), y(1), \cdots, y(N-1)]^T$$

and it can be represented as

$$\overline{\mathbf{y}} = \overline{\Lambda}\overline{\mathbf{H}}\overline{\mathbf{x}} + \overline{\mathbf{w}}, \tag{2.50}$$

where

$$\overline{\Lambda} \triangleq \text{diag}\left\{1, e^{j\Delta\omega}, \cdots, e^{j(N-1)\Delta\omega}\right\},$$

$\overline{\mathbf{H}}$ is a circulant matrix with the first column equals $[h(0), h(1), \cdots, h(N-1)]^T$, $\overline{\mathbf{x}}$ and $\overline{\mathbf{w}}$ are defined respectively as

$$\overline{\mathbf{x}} = [x(0), x(1), \cdots, x(N-1)]^T \tag{2.51}$$

$$\overline{\mathbf{w}} = [w(0), w(1), \cdots, w(N-1)]^T. \tag{2.52}$$

The matrix $\overline{\mathbf{H}}$ can be decomposed as $\mathbf{F}^H\mathbf{H}\mathbf{F}$, where $\mathbf{F}$ is a size $N \times N$ DFT matrix,

$$\mathbf{H} = \text{diag}\{H_0, H_1, \cdots, H_{N-1}\}, \tag{2.53}$$

and $H_k = \sum_{n=0}^{N-1} h(n)e^{-j\frac{2\pi kn}{N}}$. Substituting it into (2.50) and multiplying both sides of (2.50) by $\mathbf{F}$, we have

$$\mathbf{y} = \mathbf{F}\overline{\mathbf{y}} = \Lambda \mathbf{H}\mathbf{x} + \mathbf{w}, \tag{2.54}$$

$$= [Y_0, Y_1, \cdots, Y_{N-1}]^T, \tag{2.55}$$

where Λ is a circulant matrix which equals

$$\Lambda = \mathbf{F}\overline{\Lambda}\mathbf{F}^H \tag{2.56}$$

and **x** and **w** are equal to

$$\mathbf{x} = \mathbf{F}\overline{\mathbf{x}} = [X_0, X_1, \cdots, X_{N-1}]^T \tag{2.57}$$

$$\mathbf{w} = \mathbf{F}\overline{\mathbf{w}} = [W_0, W_1, \cdots, W_{N-1}]^T. \tag{2.58}$$

In (2.54), **x** still equals that in (2.33), however, due to the circulant matrix Λ, vector **y** cannot be decomposed as that in (2.35). By multiplying the circulant matrix Λ caused by the frequency offset, ICI is introduced.

2.4 The IQ Imbalance Model with Phase Noise

Ideally the carrier frequency is a pure sine wave with frequency f_c or with normalized digital frequency $\omega_c = 2\pi f_c T$, where T is the sample period. In reality, the carrier frequency may suffer from phase noise, i.e., the generated carrier frequency equals $\sin(\omega_c n + \phi_n)$ where ϕ_n is a random process. For the discussion of IQ imbalance when there is phase noise, please see [23, 24].

In free running oscillators, it has been found that the phase error ϕ_n becomes asymptotically a Brownian motion (Wiener-Levy process) as the time index $n \rightarrow \infty$. The phase noise process is given by $\phi_n = \phi_{n-1} + \varphi_n$, where the phase noise innovations φ_n are modelled as i.i.d. Gaussian random variables with 0 mean and variance σ_φ^2. Assuming perfect synchronization at the beginning of each frame, $\varphi_0 = 0$. The variance of the phase noise process increases linearly with time index n, making it non-stationary. However, the phase noise process itself is stationary and displays the so called Lorenzian spectrum. The variance of the phase noise innovations is given by

$$\sigma_\varphi^2 = 2\pi\beta T_s/N = 2\pi\beta/R \tag{2.59}$$

where T_s is the duration of an OFDM symbol, N is the number of sub-carriers, R is the symbol rate as $R \triangleq \frac{N}{T_s}$, and β is the 3 dB bandwidth of the Lorenzian spectrum for the phase noise.

Compared with the case with frequency offset, we can see that in case of phase noise, the received signal now is equal to $e^{j\phi_n} y'(n)$, where $y'(n)$ is the received baseband equivalent signal before IQ demodulation. As shown in Fig. 2.3, $y'(n)$ equals

$$\begin{aligned} y'(n) &= x'(n) \otimes h'(n) + w'(n) \\ &= x(n) \otimes \mu_t(n) \otimes h'(n) + w'(n), \end{aligned} \tag{2.60}$$

where $x(n) = s(n) + \xi_t(n) \otimes s^*(n)$.

Using approaches similar as deriving the model for the case with frequency offset, we have that after RX IQ imbalance, the received distorted signal equals

$$\begin{aligned} r(t) &= \mu_r(n) \otimes \left[e^{j\varphi_n} y'(n)\right] + \nu_r(n) \otimes \left[e^{j\varphi_n} y'(n)\right]^* \\ &= y(n) + \xi_r(n) \otimes y^*(n), \end{aligned} \tag{2.61}$$

where $y(n)$ equals to

$$\begin{aligned} y(n) &= \mu_r(n) \otimes \left[e^{j\varphi_n} y'(n)\right] \\ &= e^{j\varphi_n} \left[\left(e^{-j\varphi_n} \mu_r(n)\right) \otimes y'(n)\right]. \end{aligned} \tag{2.62}$$

Substituting (2.60) into (2.62) gives

$$\begin{aligned} y(n) &= e^{j\varphi_n} \left\{ x(n) \otimes \underbrace{\mu_t(n) \otimes h'(n) \otimes [e^{-j\varphi_n} \mu_r(n)]}_{\triangleq h(n)} + \underbrace{w'(n) \otimes [e^{-j\varphi_n} \mu_r(n)]}_{\triangleq w(n)} \right\} \\ &= e^{j\varphi_n} \{x(n) \otimes h(n) + w(n)\}. \end{aligned} \tag{2.63}$$

In frequency domain, the received signal can be written similar as Eq. (2.54), which is

$$\mathbf{y} = \Lambda \mathbf{H} \mathbf{x} + \mathbf{w}, \tag{2.64}$$

where $\mathbf{H}$, $\mathbf{x}$ and $\mathbf{w}$ are defined the same as (2.53), (2.57) and (2.58), respectively. The difference is the definition of Λ, which now is equal to

$$\Lambda = \mathbf{F} \mathrm{diag} \left\{e^{j\varphi_0}, e^{j\varphi_1}, \cdots, e^{j\varphi_{N-1}}\right\} \mathbf{F}^H. \tag{2.65}$$

2.5 The IQ Imbalance Model with Spatial Multiplexing

Spatial multiplexing or BLAST (Bell-Labs Layered Space-Time) [32] is a way to achieve high capacity in rich scattering wireless channels [32, 33]. Figure 2.5 shows the spatial multiplexing with N_t transmit and N_r receive antennas. It is also called multiple-input multiple-output (MIMO) system. For the IQ imbalance with MIMO, see for example [29, 34–36] and the references therein.

For clarity while without loss of generality, as in previous sections, we assume that an OFDM system with size N FFT/IFFT is used. We consider the transmit and

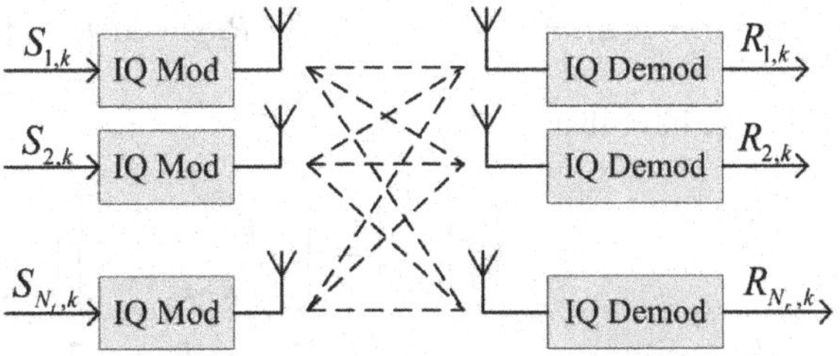

Fig. 2.5 The IQ imbalance model with MIMO spatial multiplexing

receive symbols at the k-th subcarrier. If there are no TX and RX IQ imbalances, the transmit and receive signals equal to

$$\mathbf{s}_k = [S_{1,k}, S_{2,k}, \cdots, S_{N_t,k}]^T,$$

$$\mathbf{y}_k = [Y_{1,k}, Y_{2,k}, \cdots, Y_{N_r,k}]^T.$$

They have the relation

$$\mathbf{y}_k = \mathbf{H}_k \mathbf{s}_k + \mathbf{w}_k, \tag{2.66}$$

where $\mathbf{H}_k$ is a $N_r \times N_t$ matrix, whose i-th row and j-th column $H_{i,j,k}$ is the channel coefficient between the i-th receive and j-th transmit antennas at subcarrier k. $\mathbf{w}_k$ is the $N_r \times 1$ noise vector.

If TX IQ imbalance exists, subcarrier k and $(N-k)$ interfere with each other according to Eq. (2.34). Denote the TX IQ imbalance corrupted signal as

$$\mathbf{x}_k = [X_{1,k}, X_{2,k}, \cdots, X_{N_t,k}]^T.$$

Then, after TX IQ imbalance, the signal is

$$\begin{bmatrix} \mathbf{x}_k \\ \mathbf{x}^*_{N-k} \end{bmatrix} = \begin{bmatrix} \mathbf{I}_{N_t} & \Xi_{t,k} \\ \Xi^*_{t,N-k} & \mathbf{I}_{N_t} \end{bmatrix} \begin{bmatrix} \mathbf{s}_k \\ \mathbf{s}^*_{N-k} \end{bmatrix}, \tag{2.67}$$

where $\Xi_{t,k}$ equals

$$\Xi_{t,k} = \mathrm{diag}\{\xi_{t,1,k}, \xi_{t,2,k}, \cdots, \xi_{t,N_t,k}\},$$

and $\xi_{t,n,k}$ means the IQ imbalance at the k-subcarrier of the n-th TX antenna branch.

Correspondingly, the received signal equals

$$\begin{bmatrix} \mathbf{y}_k \\ \mathbf{y}^*_{N-k} \end{bmatrix} = \begin{bmatrix} \mathbf{H}_k & \mathbf{0} \\ \mathbf{0} & \mathbf{H}^*_{N-k} \end{bmatrix} \begin{bmatrix} \mathbf{I}_{N_t} & \Xi_{t,k} \\ \Xi^*_{t,N-k} & \mathbf{I}_{N_t} \end{bmatrix} \begin{bmatrix} \mathbf{s}_k \\ \mathbf{s}^*_{N-k} \end{bmatrix} + \begin{bmatrix} \mathbf{w}_k \\ \mathbf{w}^*_{N-k} \end{bmatrix}. \tag{2.68}$$

If RX IQ imbalance exists, the received corrupted signal equals

$$\mathbf{r}_k = [R_{1,k}, R_{2,k}, \cdots, R_{N_r,k}]^T.$$

Then, we have that

$$\begin{bmatrix} \mathbf{r}_k \\ \mathbf{r}^*_{N-k} \end{bmatrix} = \begin{bmatrix} \mathbf{I}_{N_r} & \Xi_{r,k} \\ \Xi^*_{r,N-k} & \mathbf{I}_{N_r} \end{bmatrix} \begin{bmatrix} \mathbf{y}_k \\ \mathbf{y}^*_{N-k} \end{bmatrix} = \hat{\mathbf{H}}_{k,N-k} \begin{bmatrix} \mathbf{s}_k \\ \mathbf{s}^*_{N-k} \end{bmatrix} + \mathbf{v}_{k,N-k}, \tag{2.69}$$

where $\Xi_{r,k}$ equals

$$\Xi_{r,k} = \mathrm{diag}\{\xi_{r,1,k}, \xi_{r,2,k}, \cdots, \xi_{r,N_r,k}\},$$

and $\xi_{r,m,k}$ means the IQ imbalance at the k-subcarrier of the m-th RX antenna branch. $\hat{\mathbf{H}}_{k,N-k}$ equals

$$\hat{\mathbf{H}}_{k,N-k} = \begin{bmatrix} \mathbf{H}_k + \Xi_{r,k}\mathbf{H}^*_{N-k}\Xi^*_{t,N-k} & \mathbf{H}_k\Xi_{t,k} + \Xi_{r,k}\mathbf{H}^*_{N-k} \\ \Xi^*_{r,N-k}\mathbf{H}_k + \mathbf{H}^*_{N-k}\Xi^*_{t,N-k} & \Xi^*_{r,N-k}\mathbf{H}_k\Xi_{t,k} + \mathbf{H}^*_{N-k} \end{bmatrix}, \tag{2.70}$$

and $\mathbf{v}_{k,N-k}$ equals

$$\mathbf{v}_{k,N-k} = \begin{bmatrix} \mathbf{I}_{N_r} & \Xi_{r,k} \\ \Xi^*_{r,N-k} & \mathbf{I}_{N_r} \end{bmatrix} \begin{bmatrix} \mathbf{w}_k \\ \mathbf{w}^*_{N-k} \end{bmatrix}. \tag{2.71}$$

From above IQ imbalance model for spatial multiplexing systems, we can see that due to the interference between the k-th and $(N-k)$-th subcarriers, original $N_r \times N_t$ MIMO system is changed to a $2N_r \times 2N_t$ system.

2.6 The IQ Imbalance Model with Space-Time/Space-Frequency Code

Space-Time Code is a method to collect transmit spatial diversity when there are multiple antennas in transmitter [25–28]. The best known space-time code is the Alamouti's scheme [25], which works for two transmit antennas. Figure 2.6 shows

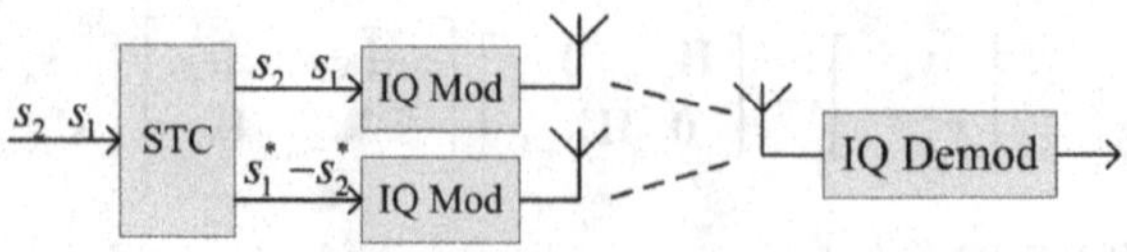

Fig. 2.6 The IQ imbalance model with space-time code/space-frequency code

the Alamouti's scheme. The impact of IQ imbalance on the space-time code is discussed in [29–31]. In this section, we follow the steps used in [30] to derive the IQ imbalance model when Alamouti's scheme is used. We assume that there are two transmit antennas and one receive antenna. For clarity, we first assume that the IQ imbalance is frequency independent, and the channel is one tap flat fading channel, then we extend the model to the frequency dependent IQ imbalance case with space-frequency code using OFDM.

For space-time code, it is assumed that the channel is constant in consecutive M time domain symbols, where $M > N_t$ and N_t is the number of transmit antennas. As described in Fig. 2.6, assume that the symbols in two consecutive time slots, i.e., s_1 and s_2, are encoded by the Alamouti's space-time code. First assume that there is no IQ imbalance at both transmitter and receiver, then after encoding, the symbols transmitted by the two transmit antennas equal to

$$\begin{bmatrix} s_1 & s_2 \\ -s_2^* & s_1^* \end{bmatrix}, \tag{2.72}$$

where row is the spatial domain and column is the time domain. Assuming that the channel coefficients are the same for the two time slots but are different for the two transmit antennas, denote them as h_1 and h_2 for the first and second transmit antennas, respectively. At the receiver, the received signals in two time slots equal to

$$\begin{bmatrix} y_1 \\ y_2 \end{bmatrix} = \begin{bmatrix} s_1 & s_2 \\ -s_2^* & s_1^* \end{bmatrix} \begin{bmatrix} h_1 \\ h_2 \end{bmatrix} + \begin{bmatrix} w_1 \\ w_2 \end{bmatrix}. \tag{2.73}$$

To do decoding, the decoupled received signal for detecting s_1 and s_2 equal to

$$\hat{y}_1 \triangleq h_1^* y_1 + h_2 y_2^* = (|h_1|^2 + |h_2|^2)\, s_1 + \hat{w}_1 \tag{2.74}$$

$$\hat{y}_2 \triangleq h_2^* y_1 - h_1 y_2^* = (|h_1|^2 + |h_2|^2)\, s_2 + \hat{w}_2, \tag{2.75}$$

where $\hat{w}_1$ and $\hat{w}_2$ are equivalent transformed noise.

If TX IQ imbalance exists, the transmitted space-time code does not equal to (2.72), but equals to the following

$$\begin{bmatrix} s_1 + \xi_{t,1} s_1^* & s_2 + \xi_{t,2} s_2^* \\ -s_2^* - \xi_{t,1} s_2 & s_1^* + \xi_{t,2} s_1 \end{bmatrix}, \tag{2.76}$$

where $\xi_{t,1}$ and $\xi_{t,2}$ are TX IQ imbalances for the first and second transmit antennas, respectively.[1]

At the receiver, if there is no IQ imbalance, the received signals now are equal to

[1]With a slight abuse of notations, here we use $\xi_{t,1}$ to denote the IQ imbalance for the first TX antennas, instead of the IQ imbalance at the first subcarrier, as is used in previous sections.

$$\begin{bmatrix} y_1 \\ y_2 \end{bmatrix} = \begin{bmatrix} s_1 + \xi_{t,1} s_1^* & s_2 + \xi_{t,2} s_2^* \\ -s_2^* - \xi_{t,1} s_2 & s_1^* + \xi_{t,2} s_1 \end{bmatrix} \begin{bmatrix} h_1 \\ h_2 \end{bmatrix} + \begin{bmatrix} w_1 \\ w_2 \end{bmatrix}. \tag{2.77}$$

If IQ imbalance exists at the receiver, assume that it is ξ_r, then the received signals at the first and second times slots are

$$\begin{bmatrix} r_1 \\ r_2 \end{bmatrix} = \begin{bmatrix} y_1 + \xi_r y_1^* \\ y_2 + \xi_r y_2^* \end{bmatrix}. \tag{2.78}$$

Substituting (2.77) into (2.78) gives us

$$\begin{aligned} r_1 &= a s_1 + b s_2 + c s_1^* + d s_2^* + v_1, \\ r_2 &= d s_1 - c s_2 + b s_1^* - a s_2^* + v_2, \end{aligned} \tag{2.79}$$

where a,b,c, and d are defined as

$$a = h_1 + \xi_r \xi_{t,1}^* h_1^*, \tag{2.80}$$

$$b = h_2 + \xi_r \xi_{t,2}^* h_2^*, \tag{2.81}$$

$$c = \xi_{t,1} h_1 + \xi_r h_1^*, \tag{2.82}$$

$$d = \xi_{t,2} h_2 + \xi_r h_2^*, \tag{2.83}$$

and v_1 are v_2 are noise terms, which equals

$$v_1 = w_1 + \xi_r w_1^*, \tag{2.84}$$

$$v_2 = w_2 + \xi_r w_2^*. \tag{2.85}$$

If OFDM is used, the frequency domain Alamouti's scheme encodes symbols over two contiguous subcarriers, instead of two contiguous time slots, and it is assumed that the channel coefficients are the same for these two contiguous subcarriers. The frequency domain Alamouti's scheme is also called space-frequency code. In this case, the encoded matrix equals

$$\begin{bmatrix} S_k & S_{k+1} \\ -S_{k+1}^* & S_k^* \end{bmatrix}. \tag{2.86}$$

IF TX IQ imbalance exists, we need to consider the space-frequency encoding at subcarrier $N-k$ and $N-k-1$, which equals

$$\begin{bmatrix} S_{N-k} & S_{N-k-1} \\ -S_{N-k-1}^* & S_{N-k}^* \end{bmatrix}. \tag{2.87}$$

According to the TX IQ imbalance interference model (2.34), assuming that IQ imbalances in consecutive subcarriers are equal, and the IQ imbalances for different transmit antennas are different, i.e. $\xi_{t,1,k} = \xi_{t,1,k+1}$ and $\xi_{t,2,k} = \xi_{t,2,k+1}$, but $\xi_{t,1,k} \neq \xi_{t,2,k}$, then the actual transmitted signal from the two transmit antennas in subcarrier k,$k+1$,$N-k$, and $N-k-1$ equal to

$$\begin{bmatrix} X_{1,k} \\ X_{1,k+1} \\ X_{1,N-k} \\ X_{1,N-k-1} \end{bmatrix} = \begin{bmatrix} 1 & 0 & \xi_{t,1,k} & 0 \\ 0 & 1 & 0 & \xi_{t,1,k} \\ \xi^*_{t,1,N-k} & 0 & 1 & 0 \\ 0 & \xi^*_{t,1,N-k} & 0 & 1 \end{bmatrix} \begin{bmatrix} S_k \\ -S^*_{k+1} \\ S^*_{N-k} \\ -S_{N-k-1} \end{bmatrix} \tag{2.88}$$

$$\begin{bmatrix} X_{2,k} \\ X_{2,k+1} \\ X_{2,N-k} \\ X_{2,N-k-1} \end{bmatrix} = \begin{bmatrix} 1 & 0 & \xi_{t,2,k} & 0 \\ 0 & 1 & 0 & \xi_{t,2,k} \\ \xi^*_{t,2,N-k} & 0 & 1 & 0 \\ 0 & \xi^*_{t,2,N-k} & 0 & 1 \end{bmatrix} \begin{bmatrix} S_{k+1} \\ S^*_{k} \\ S^*_{N-k-1} \\ S_{N-k} \end{bmatrix}. \tag{2.89}$$

At the receiver, the received signals in the four subcarriers are

$$\begin{bmatrix} Y_k \\ Y_{k+1} \end{bmatrix} = \begin{bmatrix} X_{1,k} & X_{2,k} \\ X_{1,k+1} & X_{2,k+1} \end{bmatrix} \begin{bmatrix} H_{1,k} \\ H_{2,k} \end{bmatrix} + \begin{bmatrix} W_k \\ W_{k+1} \end{bmatrix} \tag{2.90}$$

$$\begin{bmatrix} Y_{N-k} \\ Y_{N-k-1} \end{bmatrix} = \begin{bmatrix} X_{1,N-k} & X_{2,N-k} \\ X_{1,N-k-1} & X_{2,N-k-1} \end{bmatrix} \begin{bmatrix} H_{1,N-k} \\ H_{2,N-k} \end{bmatrix} + \begin{bmatrix} W_{N-k} \\ W_{N-k-1} \end{bmatrix}. \tag{2.91}$$

When there is IQ imbalances at the receiver, we also assume that $\xi_{r,k} = \xi_{r,k+1}$ and $\xi_{r,N-k} = \xi_{r,N-k-1}$, then, based on (2.37), the received signal corrupted by the RX IQ imbalance equals

$$\begin{bmatrix} R_k \\ R_{k+1} \\ R^*_{N-k} \\ R^*_{N-k-1} \end{bmatrix} = \begin{bmatrix} 1 & 0 & \xi_{r,k} & 0 \\ 0 & 1 & 0 & \xi_{r,k} \\ \xi^*_{r,N-k} & 0 & 1 & 0 \\ 0 & \xi^*_{r,N-k} & 0 & 1 \end{bmatrix} \begin{bmatrix} Y_k \\ Y_{k+1} \\ Y^*_{N-k} \\ Y^*_{N-k-1} \end{bmatrix}. \tag{2.92}$$

Reorganizing these equations, and defining

$$\mathbf{r}_k \triangleq [R_k, R_{k+1}, R^*_{N-k}, R^*_{N-k-1}]^T,$$

$$\mathbf{s}_k \triangleq [S_k, S_{k+1}, S^*_{N-k}, S^*_{N-k-1}]^T,$$

$$\mathbf{v}_k \triangleq [V_k, V_{k+1}, V^*_{N-k}, V^*_{N-k-1}]^T,$$

we have

$$\mathbf{r}_k = \mathbf{A}_k \mathbf{s}_k + \mathbf{B}_k \mathbf{s}^*_k + \mathbf{v}_k, \tag{2.93}$$

where $\mathbf{A}_k$ and $\mathbf{B}_k$ are defined as

$$\mathbf{A}_k = \begin{bmatrix} H_{1,k} & H_{1,k}\xi_{t,1,k} & H_{2,k} & H_{2,k}\xi_{t,2,k} \\ \xi_{r,k}^* H_{1,k} & \xi_{r,N-k}^* H_{1,k}\xi_{t,1,k} & \xi_{r,N-k}^* H_{2,k} & \xi_{r,N-k}^* H_{2,k}\xi_{t,2,k} \\ \xi_{r,k} H_{2,N-k}^* \xi_{t,2,N-k} & \xi_{r,k} H_{2,N-k}^* & -\xi_{r,k} H_{1,N-k}^* \xi_{t,1,N-k} & -\xi_{r,k} H_{1,N-k}^* \\ H_{2,N-k}^* \xi_{t,2,N-k} & H_{2,N-k}^* & -H_{1,N-k}^* \xi_{t,1,N-k} & -H_{1,N-k}^* \end{bmatrix}$$

$$\mathbf{B}_k = \begin{bmatrix} \xi_{r,k} H_{1,N-k}^* \xi_{t,1,N-k} & \xi_{r,k} H_{1,N-k}^* & \xi_{r,k} H_{2,N-k}^* \xi_{t,2,N-k} & \xi_{r,k} H_{2,N-k}^* \\ H_{1,N-k}^* \xi_{t,1,N-k} & H_{1,N-k}^* & H_{2,N-k}^* \xi_{t,2,N-k} & H_{2,N-k}^* \\ H_{2,k} & H_{2,k}\xi_{t,2,k} & -H_{1,k} & -H_{1,k}\xi_{t,1,k} \\ \xi_{r,N-k}^* H_{2,k} & \xi_{r,N-k}^* H_{2,k}\xi_{t,2,k} & -\xi_{r,N-k}^* H_{1,k} & -\xi_{r,N-k}^* H_{1,k}\xi_{t,1,k} \end{bmatrix},$$

and $\mathbf{v}_k$ is the colored noise, which equals

$$\mathbf{v}_k = \begin{bmatrix} 1 & 0 & \xi_{r,k} & 0 \\ 0 & 1 & 0 & \xi_{r,k} \\ \xi_{r,N-k}^* & 0 & 1 & 0 \\ 0 & \xi_{r,N-k}^* & 0 & 1 \end{bmatrix} \begin{bmatrix} W_k \\ W_{k+1} \\ W_{N-k}^* \\ W_{N-k-1}^* \end{bmatrix}. \tag{2.94}$$

If there is no TX and RX IQ imbalances, i.e., $\xi_{t,1,k} = \xi_{t,2,k} = \xi_{t,1,N-k} = \xi_{t,2,N-k} = 0$ and $\xi_{r,k} = \xi_{r,N-k} = 0$, then we have decoupled $[R_k, R_{k+1}]$ and $[R_{N-k}, R_{N-k-1}]$. By further processing, we can get decoupled S_k, S_{k+1}, S_{N-k}, and S_{N-k-1}. However, when TX and RX IQ imbalances exist, we lose these good properties.

2.7 The Impact of IQ Imbalance Model on Performance

In this section, we discuss the impact of IQ imbalance on the system performance. We mainly look at the signal-to-interference ratio (SIR) degradation caused by IQ imbalance.

From Eq. (2.34), we can see that when there is IQ imbalance, the actual transmitted signal at subcarrier k, i.e., X_k, not only includes the original signal S_k, but also includes the interference from its image subcarrier, i.e., S_{N-k}.

If there is only transmitter IQ imbalance, it is not difficult to see from Eq. (2.34) that the SIR due to transmitter IQ imbalance is

$$\mathrm{SIR}_{t,k} = \left| \frac{1}{\xi_{t,k}} \right|^2, \tag{2.95}$$

for $k = 0, 1, \cdots, N-1$. Assume that only FI IQ imbalance exists, then the SIR is the same for all subcarriers and $\xi_{t,k} = -\frac{\beta_t}{\alpha_t}$. Figure 2.7 shows how $\mathrm{SIR}_{t,k}$ changes with the transmitter IQ imbalance θ_t and ε_t when there is only FI IQ imbalance.

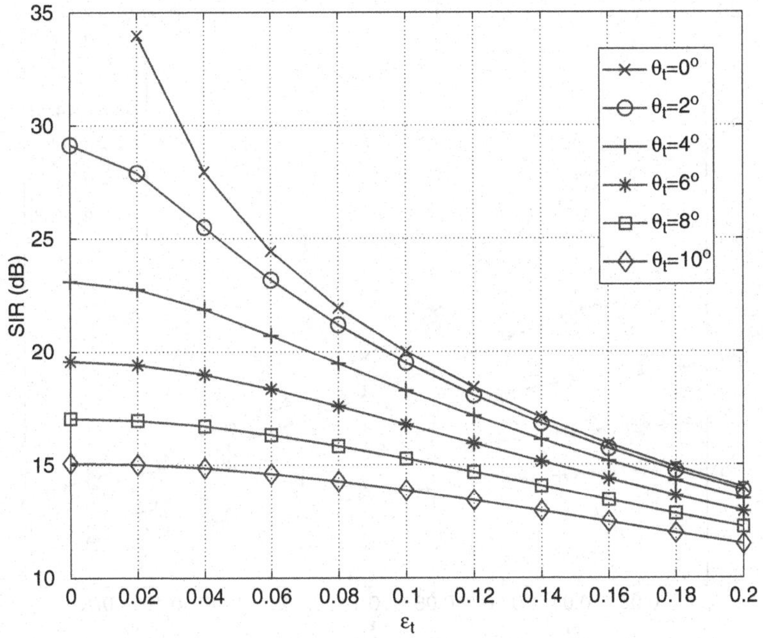

Fig. 2.7 The impact of transmitter only IQ imbalance on the SIR of the transmitted signal

Similarly, based on Eq. (2.37), we can see that when there is only receiver IQ imbalance, the SIR due to receiver IQ imbalance is

$$\mathrm{SIR}_{r,k} = \left| \frac{1}{\xi_{r,k}} \right|^2 . \tag{2.96}$$

Also, if only FI IQ imbalance exists, the SIR is the same for all subcarriers and $\xi_{r,k} = -\frac{\beta_r}{\alpha_r^*}$. Figure 2.8 shows how $\mathrm{SIR}_{r,k}$ changes with the transmitter IQ imbalance θ_r and ε_r.

If both transmitter and receiver IQ imbalances exist, then the SIR can be calculated from Eq. (2.38), which is

$$\mathrm{SIR}_k = \frac{|H_k + \xi_{r,k}\xi_{t,N-k}^* H_{N-k}^*|^2}{|\xi_{t,k} H_k + \xi_{r,k} H_{N-k}^*|^2}. \tag{2.97}$$

Let us ignore the impact of channel, i.e., assume that $H_k = 1$ for $k = 0, 1, \cdots, N-1$, then, the SIR equals

$$\mathrm{SIR}_k = \frac{|1 + \xi_{r,k}\xi_{t,N-k}^*|^2}{|\xi_{t,k} + \xi_{r,k}|^2}. \tag{2.98}$$

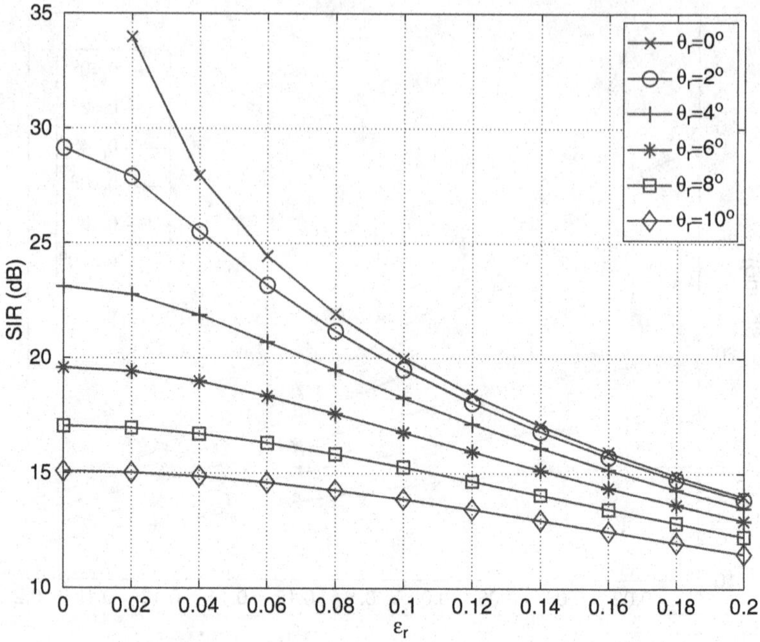

Fig. 2.8 The impact of receiver only IQ imbalance on the SIR of the received signal

If only FI IQ imbalance exists, then $\xi_{t,k} = -\frac{\beta_t}{\alpha_t}$ and $\xi_{r,k} = -\frac{\beta_r}{\alpha_r}$ for $k = 0, 1, \cdots, N-1$. In this case, the SIR equals

$$\mathrm{SIR}_k = \frac{|\alpha_r \alpha_t + \beta_r \beta_t|^2}{|\alpha_r \beta_t + \alpha_t \beta_r|^2}. \tag{2.99}$$

Figure 2.9 shows the SIR under the assumption that $\theta_t = \theta_r$ and $\epsilon_t = \epsilon_r$. Compared with the case of TX and RX only IQ imbalance, i.e., Figs. 2.7 and 2.8, we can see that there are about 6 dB degradation at the same value of phase and amplitude mismatch.

From Figs. 2.7 and 2.8 we can see that where there is only TX or RX IQ imbalance, in order to achieve 25 dB SIR, the required amplitude mismatch is about 0.04 and phase mismatch is about 2°, which is a quite stringent requirements. If both TX and RX IQ imbalances exist, at the same amplitude and phase mismatch at TX and RX, there is further 6 dB SIR degradation. This calls for the investigation of estimation and compensation of IQ imbalances.

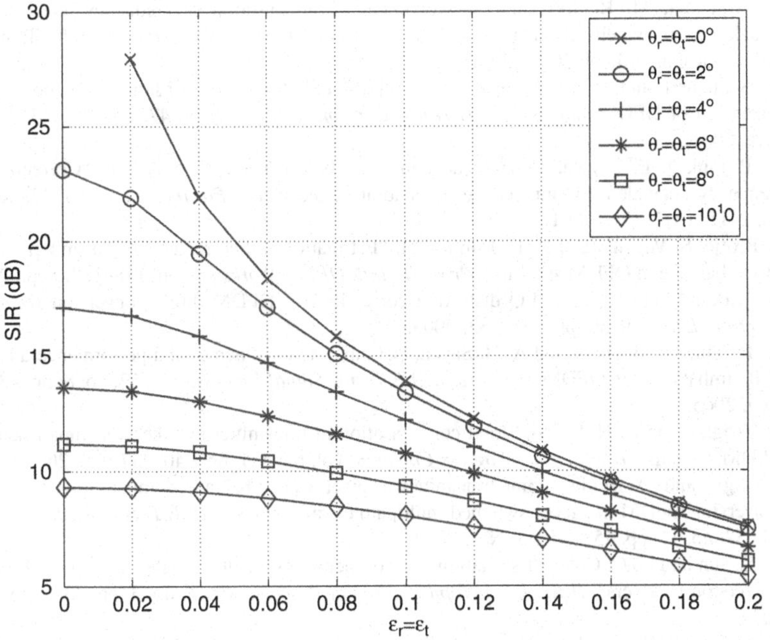

Fig. 2.9 The impact of transmitter and receiver IQ imbalance on the SIR of the received signal

2.8 Conclusions

In this chapter, we built the IQ imbalance model. We started from the general frequency dependent time domain model, and then looked at the frequency domain model. We then combined the IQ imbalance with frequency offset and phase noise. After that we discussed the IQ imbalance in the multiple antenna systems, including both spatial multiplex systems and space-time/space-frequency encoded systems. At the end, we looked at the SIR degradation due to IQ imbalance. This chapter built the foundation for the discussion of estimation and compensation of IQ imbalances in the next two chapters.

References

1. John G. Proakis, *Digital Communications*, McGraw-Hill, the Fifth Edition, 2007.
2. B. Razavi, *RF Microelectronics*, Upper Saddle River, NJ: Prentice Hall, 1998.
3. M. Valkama, M. Renfors, and V. Koivunen, "Advanced methods for I/Q imbalance compensation in communication receivers," *IEEE Trans. Signal Process.*, vol. 49, no. 10. pp. 2335–2344, Oct. 2001.

4. M. Valkama, M. Renfors, and V. Koivunen, "Compensation of frequency-selective I/Q imbalances in wideband receivers: Models and algorithms," *Proc. SPAWC 2001*, Taoyuan, Taiwan, R.O.C., March 20–23,2001
5. A. Schuchert and R. Hasholzner, "A novel IQ imbalance compensation scheme for the receiption of OFDM signals," *IEEE Trans. Consum. Electron.*, vol. 47, no. 3, pp. 313–318, Aug. 2001.
6. K. P. Pun, J. E. Franca, C. Azeredo-Leme, C. F. Chan and C. S. Choy, "Correction of frequency-dependent I/Q mismatches in quadrature receivers," *Electron. Lett.*, vol. 37, no. 23, pp. 1415–1417, Nov. 2001.
7. L. Brotje, S. Vogeler, and K.-D. Kammeyer, "Estimation and correction of transmitter-caused I/Q imbalance in OFDM systems," *Proc. 7th Intl. OFDM Workshop*, pp. 178–182, Sept. 2002.
8. W. Kirkland and K. Teo, "I/Q distortion correction for OFDM direct conversion receiver," *Electron. Lett.*, vol. 39, pp. 131–133, 2003.
9. A. Tarighat, R. Bagheri, and A. H. Sayed, "Compensation schemes and performance analysis of IQ imbalances in OFDM receivers," *IEEE Trans. Signal Process.*, vol. 53, no. 8, pp. –3268, Aug. 2005.
10. A. Tarighat and A. H. Sayed, "Joint compensation of transmitter and receiver impairments in OFDM systems," *IEEE Trans. Wireless Commun.* vol. 6, no. 1, pp. 240–247, Jan. 2007.
11. J. Feigin and D. Brady, "Joint transmitter/receiver I/Q imbalance compensation for direct conversion OFDM in packet-switched multipath environments," *IEEE Trans. Signal Process.*, vol. 57, no. 11. pp. 4588–4593, Nov. 2009.
12. K.-Y. Sun and C.-C. Chao, "Estimation and compensation of I/Q imbalance in OFDM direct-conversion receivers," *IEEE J. Sel. Topics in Signal Process.*, vol. 3, no. 3, pp. 438–453, Jun. 2009.
13. Y. Tsai, C.-P. Yen, and X. Wang, "Blind frequency-dependent I/Q imbalance compensation for direct-conversion receivers," *IEEE Trans. Wireless Commun.*, vol. 9, no. 6, pp. 1976–1986, Jun. 2010.
14. C.-J. Hsu and W.-H. Sheen, "Joint calibration of transmitter and receiver impairments in direct-conversion radio architecture," *IEEE Trans. Wireless Commun.*, vol. 11, no. 2, pp. 832–841, Feb. 2012.
15. S. Simoens, M. de Courville, F. Bourzeix, and P. de Champs, "New I/Q imbalance modeling and compensation in OFDM systems with frequency offset," *Proc. IEEE PIMRC 2002*.
16. G. Xing, M. Shen, and H. Liu, "Frequency offset and I/Q imbalance compensation for direct-conversion receivers," *IEEE Trans. Wireless Commun.*, vol. 4, no. 2, pp. 673–680, Mar. 2005.
17. D. Tandur and M. Moonen, "Joint adaptive compensation of transmitter and receiver IQ imbalance under carrier frequency offset in OFDM-based systems," *IEEE Trans. Signal Process.*, vol. 55, no. 11, pp. 5246–5252, Nov. 2007.
18. F. Horlin, A. Bourdoux, and L. V. der Perre, "Low-complexity EM-based joint acquisition of the carrier frquency offset and IQ imbalance," *IEEE Trans. Wireless Commun.*, vol. 7, no. 6, pp. 2212–2220, Jun. 2008.
19. H. Lin, X. Zhu, and K. Yamashita, "Low-complexity pilot-aided compensation for carrier frequency offset and I/Q imbalance," *IEEE Trans. Commun.*, vol. 58, no. 2, pp. 448–452, Feb. 2010.
20. Y.-H. Chung and S.-M. Phoong, "Joint estimation of I/Q imbalance, CFO and channel response for MIMO OFDM systems," *IEEE Trans. Commun.* vol. 58, no. 5, pp. 1485–1492, May 2010.
21. W. Namgoong and P. Rabiei, "CLRB-archieving I/Q mismatch estimator for low-IF receiver using repetitive training sequence in the presence of CFO," *IEEE Trans. Commun.*, vol. 60, no. 3, pp. 706–713, Mar. 2012.
22. Y.-C. Pan and S.-M. Phoon, "A time-domain joint estimation algorithm for CFO and I/Q imbalance in wideband direct-converstion receivers," *IEEE Trans. Wireless Commun.*, vol.7, no. 11, pp. 2353–2361, Nov. 2012.
23. J. Tubbax, B. Côme, L. Van der Perre, S. Donnay, M. Engels, H. D. Man, and M. Mooen, "Compensation of IQ imbalance and phase noise in OFDM systems," *IEEE Trans. on Wireless Commun.*, vol. 4, no. 3, pp. 872–877, May 2005.

24. Q. Zou, A. Tarighat, and A. H. Sayed, "Joint compensation of IQ imbalance and phase noise in OFDM wireless systems," *IEEE Trans. on Commun.*, vol. 57, no. 2, pp.404–414, Feb. 2009.
25. S. M. Alamouti, "A simple transmit diversity technique for wireless communications," *IEEE J. Select. Areas Commun.*, vol. 16, no. 8, pp. 1451–1458, Oct. 1998.
26. V. Tarokh, N. Seshadri, and A. R. Calderbank, "Space-Time codes for high data rate wireless communication: performance criterion and code construction," *IEEE Trans. on Inform. Theory*, vol.44, no. 2, pp. 744–765, Mar. 1998.
27. V. Tarokh, H. Jafarkhani and A. R. Calderbank, "Space-time block codes for orthogonal designs," *IEEE Trans. on Inform. Theory*, vol.45, no. 5, pp. 1456–1467, July. 1999.
28. Haiquan Wang and Xiang-Gen Xia, "Upper bounds of rates of complex orthogonal space-time block codes," *IEEE Trans. on Inform. Theory*, vol. 49, no. 10, pp. 2788–2796, Oct. 2003.
29. A. Tarighat and A. H. Sayed, "MIMO OFDM receivers for systems with IQ imbalances," *IEEE Trans. Signal Process.*, vol. 53, no. 9, pp. 3583–3596, Sept. 2005.
30. Y. Zou, M. Valkama, and M. Renfors, "Digital compensation of I/Q imbalance effects in space-time coded transmit diversity systems," *IEEE Trans. Signal Process.*, vol. 56, no. 6, pp. 2496–2508, Jun. 2008.
31. J. Qi and S. Aïssa, "Analysis and compensation of I/Q imbalance in MIMO transmit-receive diversity systems," *IEEE Trans. Commun.*, vol. 58, no. 5, pp. 1546–1556, May 2010.
32. G. L. Foschini and M. J. Gans, "On limits of wireless communications in a fading environment when using multiple antennas", *Wireless Personal Communications*, vol. 6, pp. 311–335, Mar. 1998.
33. E. Telatar, "Capacity of multi-antenna Gaussian channels," *AT&T Bell-Labs, Tech. Rep.*, June. 1995.
34. T. Schenk, P. Smulders, and E. Fledderus, "Estimation and compensation of TX and RX IQ imbalance in OFDM-based MIMO systems," *Proc. IEEE Radio and Wireless Symposium*, pp. 215–218, 2006.
35. H. Minn and D. Munoz, "Pilot designs for channel estimation of MIMO OFDM systems with frequency-depedent I/Q imbalances," *IEEE Trans. Commun.*, vol. 58, no. 8, pp. 2252–2264, Aug. 2010.
36. J. Luo, W. Keusgen, and A. Kortke, "Preamble designs for efficient joint channel and frequency-selective I/Q-imbalance compensation in MIMO-OFDM systems," *Proc. IEEE WCNC 2010.*

Chapter 3
Frequency Independent IQ Imbalance Estimation and Compensation

Abstract For narrow band systems, the IQ imbalance can be assumed to be frequency independent. In this chapter, we look at the frequency independent IQ imbalance estimation and compensation. We start from simple RX only IQ imbalance estimation and compensation, then discuss joint TX and RX IQ imbalance compensation and estimation. After that we discuss the estimation and compensation of IQ imbalance when there is frequency offset. At last, we look at the IQ imbalance for multiple antenna systems.

3.1 Introduction

Though the IQ imbalance is frequency dependent (FD) in nature, for narrow band systems, the IQ imbalance can be approximated as frequency independent (FI). In this case, based on the model derived in last chapter, no filtering operation is involved, and for TX (or RX), there are only two coefficients, i.e., amplitude mismatch ε_t (or ε_r) and phase mismatch θ_t (or θ_r). The compensation and estimation for this case is simpler than those for FD IQ imbalance. Major existing literatures about IQ imbalance are based on the assumption of FI IQ imbalance, see for example [1–28] and the references therein.

For standardized communication systems, because of the requirement of interoperability, the TX IQ imbalance should be designed to meet the error vector magnitude (EVM) requirement specified in the standard. So, when designing a system, RX only IQ imbalance is a major issue. For non-standardized communication system, the requirement of TX IQ imbalance can be relaxed, and at the receiver, joint TX and RX IQ imbalance compensation and estimation can be used to lower the cost of overall systems. In this chapter, we first look at the simple RX only IQ imbalance compensation and estimation, then look at the joint TX and RX IQ imbalance compensation and estimation. After that we discuss the compensation and estimation of IQ imbalance when there is frequency offset. At last, we look at the IQ imbalance when there are multiple transmit and receive antennas.

Y. Li, *In-Phase and Quadrature Imbalance: Modeling, Estimation, and Compensation*, SpringerBriefs in Electrical and Computer Engineering,
DOI 10.1007/978-1-4614-8618-3_3,

3.2 RX Only IQ Imbalance Estimation and Compensation Without Other Impairments

As derived in Eq. (2.30), when only FI RX side IQ imbalance exists, in time domain, there is $\xi_t(n) = 0$, and $\xi_r(n) = -\frac{\beta_r}{\alpha_r^*} \triangleq \xi_r$. The transmitted signal $s(n)$ and the received signal $r(n)$ have relation

$$r(n) = \underbrace{[h(n) \otimes s(n) + w(n)]}_{\triangleq y(n)} + \xi_r \underbrace{[h(n) \otimes s(n) + w(n)]^*}_{= y^*(n)}. \tag{3.1}$$

In frequency domain, based on Eq. (2.38), the model can be written as

$$\underbrace{\begin{bmatrix} R_k \\ R_{N-k}^* \end{bmatrix}}_{\triangleq \mathbf{r}_k} = \underbrace{\begin{bmatrix} H_k & \xi_r H_{N-k}^* \\ \xi_r^* H_k & H_{N-k}^* \end{bmatrix}}_{\triangleq \mathbf{H}_k} \underbrace{\begin{bmatrix} S_k \\ S_{N-k}^* \end{bmatrix}}_{\triangleq \mathbf{s}_k} + \underbrace{\begin{bmatrix} 1 & \xi_r \\ \xi_r^* & 1 \end{bmatrix} \begin{bmatrix} W_k \\ W_{N-k}^* \end{bmatrix}}_{\triangleq \mathbf{v}_k}. \tag{3.2}$$

3.2.1 Compensation

Assume that the value of ξ_r is known, then in time domain, using (3.1), we can get

$$\begin{aligned} \hat{r}(n) &= r(n) - \xi_r r^*(n) \\ &= (1 - |\xi_r|^2) y(n). \end{aligned} \tag{3.3}$$

This means that we can process the received signal $r(n)$ as

$$\begin{aligned} y(n) &= \frac{r(n) - \xi_r r^*(n)}{1 - |\xi_r|^2} \\ &= h(n) \otimes s(n) + w(n), \end{aligned} \tag{3.4}$$

and then use conventional time domain equalization to detect the transmitted symbol. This is equivalent to the pre-FFT distortion correction method described in [6].

In the frequency domain, if ξ_r is known, using (3.2), we can do

$$\begin{aligned} \begin{bmatrix} Y_k \\ Y_{N-k}^* \end{bmatrix} &= \begin{bmatrix} 1 & \xi_r \\ \xi_r^* & 1 \end{bmatrix}^{-1} \begin{bmatrix} R_k \\ R_{N-k}^* \end{bmatrix} \\ &= \begin{bmatrix} H_k & 0 \\ 0 & H_{N-k}^* \end{bmatrix} \begin{bmatrix} S_k \\ S_{N-k}^* \end{bmatrix} + \begin{bmatrix} W_k \\ W_{N-k}^* \end{bmatrix}. \end{aligned} \tag{3.5}$$

This implies that the detection of S_k and S_{N-k} is decoupled, and the conventional single tap equalizer for OFDM can be used.

If ξ_r is not known, but the equivalent channel matrix $\mathbf{H}_k$, defined in (3.2), is known, then, in the frequency domain, regularized least square (LS) method [6] can be used to estimate the transmitted tone pair $\mathbf{s}_k$, i.e.,

$$\tilde{\mathbf{s}}_k = (\mathbf{H}_k^H \mathbf{H}_k + \delta \mathbf{I}_2)^{-1} \mathbf{H}_k^H \mathbf{r}_k, \tag{3.6}$$

where δ is a regularization parameter.

Based on the received symbol vector $\mathbf{r}_k$, [6] proposed another method to recover $\mathbf{s}_k$ using adaptive filter. In this case, at time instant m (the m-th OFDM symbol, or, equivalently, the m-th iteration), assuming that the 2×1 weight vectors to recover $S_{k,m}$ and $S^*_{N-k,m}$ in $\mathbf{s}_{k,m}$ are $\mathbf{a}_{k,m}$ and $\mathbf{a}_{N-k,m}$, respectively, then we have

$$\tilde{S}_{k,m} = \mathbf{a}_{k,m}^H \mathbf{r}_{k,m}, \tag{3.7}$$

$$\tilde{S}^*_{N-k,m} = \mathbf{a}_{N-k,m}^H \mathbf{r}_{k,m}. \tag{3.8}$$

If Least Mean Square (LMS) adaptive filter is used, $\mathbf{a}_{k,m}$ and $\mathbf{a}_{N-k,m}$ can be updated as

$$\mathbf{a}_{k,m+1} = \mathbf{a}_{k,m} + \mu_{LMS}\, \mathbf{r}^*_{k,m}\, e_{k,m}, \tag{3.9}$$

$$\mathbf{a}_{N-k,m+1} = \mathbf{a}_{N-k,m} + \mu_{LMS}\, \mathbf{r}^*_{k,m}\, e_{N-k,m}, \tag{3.10}$$

where $e_{k,m} = S_{k,m} - \mathbf{a}_{k,m}^H \mathbf{r}_{k,m}$ and $e_{N-k,m} = S^*_{N-k,m} - \mathbf{a}_{N-k,m}^H \mathbf{r}_{k,m}$, and $S_{k,m}$ and $S_{N-k,m}$ are known training symbols. μ_{LMS} is the step size for the iteration. To speed up the convergence, at the beginning of the iteration, we can set the weight vector to be the estimated channel under the assumption that there is no IQ imbalance, i.e, setting

$$\mathbf{a}_{k,0} = [\tilde{H}_k \;\; 0]^T \tag{3.11}$$

$$\mathbf{a}_{N-k,0} = [0 \;\; \tilde{H}^*_{N-k}], \tag{3.12}$$

where

$$\tilde{H}_k = \frac{\sum_{m=1}^{M_{tr}} S^*_{k,m} R_{k,m}}{\sum_{m=1}^{M_{tr}} S^*_{k,m} S_{k,m}}, \tag{3.13}$$

and M_{tr} is the total number of training symbols.

3.2.2 Parameter Estimation

To estimate $\mathbf{H}_k$ in (3.2), i.e., the combined IQ imbalance and channel, if it is assumed that the wireless channel keeps constant in M_{tr} OFDM symbols, writing the received tone pair $\mathbf{r}_k$ in the m-th OFDM symbol as $\mathbf{r}_{k,m}$ and the transmitted signal in the k-th subcarrier of the m-th OFDM symbol as $S_{k,m}$, then similar as in [6], based on (3.2), we have

$$\mathbf{r}_{k,m} = \underbrace{\begin{bmatrix} S_{k,m} & 0 & S^*_{N-k,m} & 0 \\ 0 & S_{k,m} & 0 & S^*_{N-k,m} \end{bmatrix}}_{\triangleq \mathbf{S}_{k,m}} \times \underbrace{\begin{bmatrix} H_k \\ \xi_r^* H_k \\ \xi_r H^*_{N-k} \\ H^*_{N-k} \end{bmatrix}}_{\triangleq \mathbf{h}_k} + \mathbf{v}_{k,m}. \tag{3.14}$$

Stack the received tone pair in M_{tr} OFDM symbols, we can get

$$\underbrace{\begin{bmatrix} \mathbf{r}_{k,1} \\ \mathbf{r}_{k,2} \\ \vdots \\ \mathbf{r}_{k,M_{tr}} \end{bmatrix}}_{\triangleq \hat{\mathbf{r}}_k} = \underbrace{\begin{bmatrix} \mathbf{S}_{k,1} \\ \mathbf{S}_{k,2} \\ \vdots \\ \mathbf{S}_{k,M_{tr}} \end{bmatrix}}_{\triangleq \mathbf{S}_k} \times \mathbf{h}_k + \underbrace{\begin{bmatrix} \mathbf{v}_{k,1} \\ \mathbf{v}_{k,2} \\ \vdots \\ \mathbf{v}_{k,M_{tr}} \end{bmatrix}}_{\triangleq \hat{\mathbf{v}}_k}. \tag{3.15}$$

The estimation of $\mathbf{h}_k$ can be written as

$$\tilde{\mathbf{h}}_k = (\mathbf{S}_k^H \mathbf{S}_k + \delta \mathbf{I}_4)^{-1} \mathbf{S}_k^H \hat{\mathbf{r}}_k. \tag{3.16}$$

$\mathbf{S}_k$ can be specially designed training symbols, as shown in [29]. If training symbols are not available, decision feedback low order modulation symbols can be used for this estimation purpose.

If it is desired to estimate ξ_r, having the estimated value $\tilde{\mathbf{h}}_k \triangleq [\tilde{h}_{k,1}, \tilde{h}_{k,2}, \tilde{h}_{k,3}, \tilde{h}_{k,4}]^T$, based on the definition of $\mathbf{h}_k$ shown in (3.14), ξ_r can be estimated as [6]

$$\tilde{\xi}_r = \frac{1}{N-2} \sum_{k=1}^{N/2-1} \left[\left(\frac{\tilde{h}_{k,2}}{\tilde{h}_{k,1}} \right)^* + \left(\frac{\tilde{h}_{k,3}}{\tilde{h}_{k,4}} \right) \right]. \tag{3.17}$$

To simplify the estimation, training symbols like those having the pattern as shown in Fig. 3.1 can be used [6]. Using this pattern, for $m = 1, 2, \cdots, M_{tr}/2$ and $k = 1, 2, \cdots, N/2 - 1$, we have

$$R_{k,m} = H_k S_{k,m} + V_{k,m}$$
$$R^*_{N-k,m} = \xi_r^* H_k S_{k,m} + V^*_{N-k,m}.$$

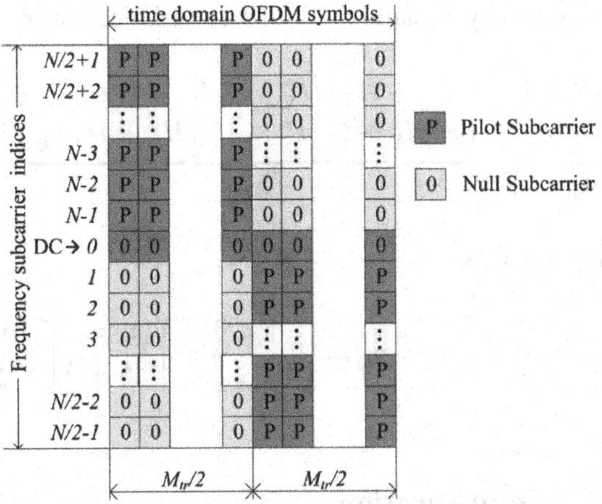

Fig. 3.1 The training symbols to estimate the combined IQ imbalance parameter and channel

Similarly, for $m = M_{tr}+1, M_{tr}+2, \cdots, M_{tr}$ and $k = 1, 2, \cdots, N/2-1$, we have

$$\begin{aligned} R_{k,m} &= \xi_r H^*_{N-k} S^*_{N-k,m} + V_{k,m} \\ R^*_{N-k,m} &= H^*_{N-k} S^*_{N-k,m} + V^*_{N-k,m}. \end{aligned}$$

To reduce the impact of noise, we can do average over $M_{tr}/2$ OFDM symbols, then the estimation of ξ_r can be written as

$$\tilde{\xi}_r = \frac{1}{N-2} \sum_{k=1}^{N/2-1} \left[\frac{\sum_{m=1}^{M_{tr}/2} (R^*_{N-k,m}/S_{k,m})}{\sum_{m=1}^{M_{tr}/2} (R_{k,m}/S_{k,m})} + \frac{\sum_{m=M_{tr}/2+1}^{M_{tr}} (R_{k,m}/S^*_{N-k,m})}{\sum_{m=M_{tr}/2+1}^{M_{tr}} (R^*_{N-k,m}/S^*_{N-k,m})} \right]. \tag{3.18}$$

3.3 Joint TX and RX IQ Imbalance Estimation and Compensation Without Other Impairments

If both transmitter and receiver IQ imbalances exist, and both of them are FI, i.e., $\xi_t(n) = -\frac{\beta_t}{\alpha_t} \triangleq \xi_t$ and $\xi_r(n) = -\frac{\beta_r}{\alpha_r^*} \triangleq \xi_r$, in the time domain, the transmitted and received signal have relation according to Fig. 2.4. The TX IQ imbalance corrupted transmitted signal $x(n)$ equals

$$x(n) = s(n) + \xi_t\, s^*(n). \tag{3.19}$$

The RX IQ imbalance corrupted received signal can be written as

$$r(n) = \underbrace{h(n) \otimes x(n) + w(n)}_{\triangleq y(n)} + \xi_r \underbrace{[h(n) \otimes x(n) + w(n)]^*}_{= y^*(n)} \tag{3.20}$$

In frequency domain, the k-th received tone pair $\mathbf{r}_k$ equal to

$$\underbrace{\begin{bmatrix} R_k \\ R^*_{N-k} \end{bmatrix}}_{\triangleq \mathbf{r}_k} = \underbrace{\begin{bmatrix} H_k + \xi_r \xi_t^* H^*_{N-k} & \xi_t H_k + \xi_r H^*_{N-k} \\ \xi_r^* H_k + \xi_t^* H^*_{N-k} & \xi_r^* \xi_t H_k + H^*_{N-k} \end{bmatrix}}_{\triangleq \mathbf{H}_k} \underbrace{\begin{bmatrix} S_k \\ S^*_{N-k} \end{bmatrix}}_{\triangleq \mathbf{s}_k} + \underbrace{\begin{bmatrix} V_k \\ V^*_{N-k} \end{bmatrix}}_{\triangleq \mathbf{v}_k}, \quad (3.21)$$

where $\mathbf{v}_k$ equals

$$\mathbf{v}_k = \begin{bmatrix} V_k \\ V^*_{N-k} \end{bmatrix} = \begin{bmatrix} 1 & \xi_r \\ \xi_r^* & 1 \end{bmatrix} \begin{bmatrix} W_k \\ W^*_{N-k} \end{bmatrix}. \quad (3.22)$$

3.3.1 Compensation

Assuming that ξ_t and ξ_r are known by the transmitter and receiver, respectively. In time domain, at the transmitter, in stead of sending $s(n)$ directly, we can send

$$\hat{s}(n) = \frac{s(n) - \xi_t s^*(n)}{1 - |\xi_t|^2}. \quad (3.23)$$

Then after distortion by the TX IQ imbalance, the actual transmitted signal is the uncontaminated signal $s(n)$, similar as the pre-FFT approach described in [26].

At the receiver, the compensation can be carried out similarly as that in Eq. (3.4).

In frequency domain, if ξ_t is known by the transmitter, then instead of sending $\mathbf{s}_k$ directly, the signal $\hat{\mathbf{s}}_k$ defined below can be sent,

$$\hat{\mathbf{s}}_k = \begin{bmatrix} 1 & \xi_t \\ \xi_t^* & 1 \end{bmatrix}^{-1} \mathbf{s}_k. \quad (3.24)$$

At the receiver, using the same equation as (3.5), we can remove the impact of RX IQ imbalance, and simple single tap equalization can be used to detect the transmitted symbol.

If exact TX and RX IQ imbalances are not known, however, the combined channel $\mathbf{H}_k$ defined in (3.21) is known, then, similar as Eq. (3.6), the transmitted symbol can be estimated as [26]

$$\tilde{\mathbf{s}}_k = (\mathbf{H}_k^H \mathbf{H}_k + \delta \mathbf{I}_2)^{-1} \mathbf{H}_k^H \mathbf{r}_k, \quad (3.25)$$

where δ is a regularization parameter.

Also, adaptive filter can be used to recover the transmitted symbol $\mathbf{s}_k$ from the received symbol $\mathbf{r}_k$ [26]. Assuming that M_{tr} training symbols are used, then the adaptive filtering operations can be carried out the same as those in Eqs. (3.7) and (3.8), and the filter coefficients can be updated the same as those in Eqs. (3.9) and (3.10).

To speed up the convergence, at the beginning, assuming that $\xi_t \approx 0$ and $\xi_r \approx 0$, we can set the initial weights the same as (3.11) and (3.12), and the initial estimated channel as (3.13).

3.3.2 Parameter Estimation

Besides adaptive filtering method, other compensation methods need either the combined effect of channel and TX and IQ imbalances, or only TX and RX IQ imbalances.

To estimate the combined channel and TX and RX IQ imbalances, assuming that M_{tr} OFDM symbols are used for training and the channel keeps constant during these symbols, based on (3.21), we can write the k-th received tone pair in the m-th OFDM symbol time as $\mathbf{r}_{k,m}$, which equals [26]

$$\mathbf{r}_{k,m} = \underbrace{\begin{bmatrix} S_{k,m} & 0 & S^*_{N-k,m} & 0 \\ 0 & S_{k,m} & 0 & S^*_{N-k,m} \end{bmatrix}}_{\triangleq \mathbf{S}_{k,m}} \times \underbrace{\begin{bmatrix} H_k + \xi_r \xi_t^* H^*_{N-k} \\ \xi_r^* H_k + \xi_t^* H^*_{N-k} \\ \xi_t H_k + \xi_r H^*_{N-k} \\ \xi_r^* \xi_t H_k + H^*_{N-k} \end{bmatrix}}_{\triangleq \mathbf{h}_k} + \mathbf{v}_{k,m}. \quad (3.26)$$

Compared with (3.14), we can see that the difference is only in the definition of the equivalent channel $\mathbf{h}_k$. So, by using the same method described in Sect. 3.2.2, we can get the estimation $\tilde{\mathbf{h}}_k$, and correspondingly the estimation of $\mathbf{H}_k$ [26].

Using the pilot pattern in Fig. 3.1, the estimation can be simplified. Assuming that $\tilde{\mathbf{h}}_k = [\tilde{h}_{k,1}, \tilde{h}_{k,2}, \tilde{h}_{k,3}, \tilde{h}_{k,4}]$, the components of $\mathbf{h}_k$ can be estimated individually, which equals

$$\tilde{h}_{k,1} = \frac{2}{M_{tr}} \sum_{m=1}^{M_{tr}/2} R_{k,m}/S_{k,m}, \quad (3.27)$$

$$\tilde{h}_{k,2} = \frac{2}{M_{tr}} \sum_{m=1}^{M_{tr}/2} R_{N-k,m}/S_{k,m}, \quad (3.28)$$

$$\tilde{h}_{k,3} = \frac{2}{M_{tr}} \sum_{m=M_{tr}/2+1}^{M_{tr}} R_{k,m}/S^*_{N-k,m}, \quad (3.29)$$

$$\tilde{h}_{k,4} = \frac{2}{M_{tr}} \sum_{m=M_{tr}/2+1}^{M_{tr}} R_{N-k,m}/S^*_{N-k,m}. \quad (3.30)$$

In case it is desired to obtain the TX and RX IQ imbalances, i.e., ξ_t and ξ_r. If $\mathbf{h}_k$ is known, based on the structure of $\mathbf{h}_k$, under the assumption that $|\xi_r \xi_t^*| \ll 1$, we can approximate the channel coefficient at subcarrier k and $N-k$, i.e., H_k and H_{N-k}, respectively as

$$\tilde{H}_k = \tilde{h}_{k,1}, \quad \tilde{H}_{N-k} = \tilde{h}^*_{k,4}. \tag{3.31}$$

Then, ξ_t and ξ_r can be estimated as

$$\begin{bmatrix} \tilde{\xi}_t \\ \tilde{\xi}_r \end{bmatrix} = \frac{2}{N-2} \sum_{k=1}^{N/2-1} \begin{bmatrix} \tilde{h}_{k,1} & \tilde{h}_{k,4} \\ \tilde{h}^*_{k,4} & \tilde{h}^*_{k,1} \end{bmatrix}^{-1} \begin{bmatrix} \tilde{h}_{k,3} \\ \tilde{h}^*_{k,2} \end{bmatrix}. \tag{3.32}$$

Another approach to estimate the TX and RX IQ imbalances and the channel is from the time domain. In the following, using the method described in [19], we show how to get the TX and RX IQ imbalances and the channel estimates from the time domain.

We assume that CP with sufficient length is used in the system, and the system can be either OFDM or SC-FDE. After removing CP, the time domain convolution in Eq. (3.20) can be written in matrix form as

$$\overline{\mathbf{r}} = \underbrace{[\overline{\mathbf{X}}\,\overline{\mathbf{h}} + \overline{\mathbf{w}}]}_{\triangleq \overline{\mathbf{y}}} + \xi_r \underbrace{[\overline{\mathbf{X}}\,\overline{\mathbf{h}} + \overline{\mathbf{w}}]^*}_{= \overline{\mathbf{y}}^*}, \tag{3.33}$$

where

$$\overline{\mathbf{r}} = [r(0), r(1), \cdots, r(N-1)]^T$$
$$\overline{\mathbf{h}} = [h(0), h(1), \cdots, h(L_h - 1), \mathbf{0}_{N-L_h}]^T$$
$$\overline{\mathbf{w}} = [w(0), w(1), \cdots, w(N-1)]$$

and $\overline{\mathbf{X}}$ is an $N \times N$ circulant matrix with the first column defined as

$$\overline{\mathbf{x}} = [x(0), x(1), \cdots, x(N-1)]^T.$$

Based on (3.19), $\overline{\mathbf{X}}$ can be written as

$$\overline{\mathbf{X}} = \overline{\mathbf{S}} + \xi_t \overline{\mathbf{S}}^*, \tag{3.34}$$

where $\overline{\mathbf{S}}$ is a circulant matrix with the first column defined as

$$\overline{\mathbf{s}} = [s(0), s(1), \cdots, s(N-1)]^T.$$

Assuming that ξ_t and ξ_r are known, then according to the compensation method in (3.4), the RX IQ imbalance compensated signal can be written as

$$\overline{\mathbf{y}} = \frac{\overline{\mathbf{r}} - \xi_r \overline{\mathbf{r}}^*}{1 - |\xi_r|^2} = \overline{\mathbf{X}}\,\overline{\mathbf{h}} + \overline{\mathbf{w}}. \tag{3.35}$$

The LS estimation of $\overline{\mathbf{h}}$ can be estimated as

$$\tilde{\overline{\mathbf{h}}} = \overline{\mathbf{X}}^{-1}\overline{\mathbf{y}} = (\overline{\mathbf{S}} + \xi_t \overline{\mathbf{S}}^*)^{-1}\overline{\mathbf{y}}, \tag{3.36}$$

where $\tilde{\overline{\mathbf{h}}} = [\tilde{h}(0), \tilde{h}(1), \cdots, \tilde{h}(N-1)]^T$.

Ideally, if no TX and RX IQ imbalances exist, the $(L_h + 1)$-th to the $(N - 1)$-th components in $\tilde{\overline{\mathbf{h}}}$ are affected only by noise. As observed, the TX and RX IQ imbalance may increase the signal level for the $(L_h + 1)$-th to the $(N - 1)$-th components in $\tilde{\overline{\mathbf{h}}}$. Therefore, the TX and RX IQ imbalance ξ_t and ξ_r can be estimated to minimize the sum of the energy of the $(L_h + 1)$-th to the $(N - 1)$-th components in $\tilde{\overline{\mathbf{h}}}$. Defining the channel residue energy (CRE) [19] as

$$\text{CRE} = \sum_{n=L_h+1}^{N-1} |\tilde{h}(n)|^2 = \left\| \mathbf{P}\tilde{\overline{\mathbf{h}}} \right\|^2, \tag{3.37}$$

where $\mathbf{P} = [\mathbf{0} \quad \mathbf{I}_{N-L_h-1}]$ is an $(N - L_h - 1) \times N$ matrix.

By some manipulations, the CRE can be written as

$$\text{CRE} = \left\| \mathbf{P}(\mathbf{I}_N + \xi_t \overline{\mathbf{S}}^{-1}\overline{\mathbf{S}}^*)^{-1}\overline{\mathbf{S}}^{-1} \frac{\overline{\mathbf{r}} - \xi_r \overline{\mathbf{r}}^*}{1 - |\xi_r|^2} \right\|^2. \tag{3.38}$$

Assume that $\xi_t \ll 1$ and $\xi_r \ll 1$, we can do approximations: $1 - |\xi_r|^2 \approx 1$, $(\mathbf{I}_N + \xi_t \overline{\mathbf{S}}^{-1}\overline{\mathbf{S}}^*)^{-1} \approx \mathbf{I}_N - \xi_t \overline{\mathbf{S}}^{-1}\overline{\mathbf{S}}^*$, and $\xi_t \xi_r \approx 0$. Having these, the CRE can be approximated as

$$\text{CRE} = \left\| \mathbf{P}\overline{\mathbf{S}}^{-1}\overline{\mathbf{r}} - \Phi \begin{bmatrix} \xi_r \\ \xi_t \end{bmatrix} \right\|^2, \tag{3.39}$$

where Φ is defined as

$$\Phi = \left[\mathbf{P}\overline{\mathbf{S}}^{-1}\overline{\mathbf{r}}^* \quad \mathbf{P}\overline{\mathbf{S}}^{-1}\overline{\mathbf{S}}^*\overline{\mathbf{S}}^{-1}\overline{\mathbf{r}} \right]. \tag{3.40}$$

Then, ξ_r and ξ_t can be estimated as

$$\begin{bmatrix} \tilde{\xi}_r \\ \tilde{\xi}_t \end{bmatrix} = (\Phi^H \Phi)^{-1}\Phi^H \mathbf{P}\overline{\mathbf{S}}^{-1}\overline{\mathbf{r}}. \tag{3.41}$$

Compared with (3.32), we can see that the estimation using (3.41) needs to know the channel length L_h, and needs to handle an $(N - L_h - 1) \times (N - L_h - 1)$ matrices. If the channel length L_h is short, the complexity is higher than that using (3.32).

Once obtaining the estimations of ξ_r and ξ_t, we can substitute them into (3.36) to get the estimation of the channel $\overline{\mathbf{h}}$.

3.4 Joint TX and RX IQ Imbalance Estimation and Compensation in Presence of Frequency Offset

When there are frequency offset and frequency independent TX and RX IQ imbalances, starting from Eqs. (2.46), (2.48) and (2.49), and setting $\xi_t(n) = -\frac{\beta_t}{\alpha_t} \triangleq \xi_t$ and $\xi_r(n) = -\frac{\beta_r}{\alpha_r^*} \triangleq \xi_r$, we have

$$\begin{aligned} r(n) &= e^{j\Delta\omega n}[x(n) \otimes h(n) + w(n)] + \xi_r e^{-j\Delta\omega n}[x^*(n) \otimes h^*(n) + w^*(n)] \\ &= e^{j\Delta\omega n}[s(n) \otimes h(n)] + \xi_t e^{j\Delta\omega n}[s^*(n) \otimes h(n)] \\ &\quad + \xi_r e^{-j\Delta\omega n}[s^*(n) \otimes h^*(n)] + \xi_r \xi_t^* e^{-j\Delta\omega n}[s(n) \otimes h^*(n)] \\ &\quad + v(n), \end{aligned} \tag{3.42}$$

where $v(n)$ is the noise item, which equals

$$v(n) = e^{j\Delta\omega n} w(n) + \xi_r e^{-j\Delta\omega n} w^*(n). \tag{3.43}$$

Assuming that the CP is sufficiently long, after removing CP at the receiver, we can write the time domain equation in matrix form, which is

$$\begin{aligned} \overline{\mathbf{r}} &= \overline{\Lambda}\,\overline{\mathbf{S}}\,\overline{\mathbf{h}} + \xi_t \overline{\Lambda}\,\overline{\mathbf{S}}^*\overline{\mathbf{h}} + \xi_r \overline{\Lambda}^{-1}\overline{\mathbf{S}}^*\overline{\mathbf{h}}^* + \xi_r \xi_t^* \overline{\Lambda}^{-1}\overline{\mathbf{S}}\,\overline{\mathbf{h}}^* + \overline{\mathbf{v}} \\ &= \overline{\Lambda}\,\underbrace{[\overline{\mathbf{S}}\,\overline{\mathbf{h}} + \xi_t \overline{\mathbf{S}}^*\overline{\mathbf{h}}]}_{\triangleq \overline{\mathbf{p}}} + \xi_r \overline{\Lambda}^{-1}\underbrace{[\overline{\mathbf{S}}\,\overline{\mathbf{h}} + \xi_t \overline{\mathbf{S}}^*\overline{\mathbf{h}}]^*}_{= \overline{\mathbf{p}}^*} + \overline{\mathbf{v}}, \end{aligned} \tag{3.44}$$

where

$$\begin{aligned} \overline{\Lambda} &= \mathrm{diag}\{1, e^{j\Delta\omega}, \cdots, e^{j\Delta\omega n}\} \\ \overline{\mathbf{h}} &= [h(0), h(1), \cdots, h(L_h - 1), \mathbf{0}_{N-L_h}]^T \\ \overline{\mathbf{v}} &= [v(0), v(1), \cdots, v(N-1)] \end{aligned}$$

and $\overline{\mathbf{S}}$ is a circulant matrix with the first column defined as

$$\overline{\mathbf{s}} = [s(0), s(1), \cdots, s(N-1)]^T.$$

3.4.1 Compensation

Assume that ξ_t is known at the transmitter and the frequency offset $\Delta\omega$ and RX IQ imbalances ξ_r are known at the receiver, we can compensate the received signal $\overline{\mathbf{r}}$ in the following steps to recover the transmitted signal $\overline{\mathbf{s}}$.

The first step is to do preprocessing to compensate the TX IQ imbalance at the transmitter, i.e, instead of sending $\overline{\mathbf{s}}$, we send

$$\hat{\overline{\mathbf{s}}} = \frac{\overline{\mathbf{s}} - \xi_t \overline{\mathbf{s}}^*}{1 - |\xi_t|^2}. \tag{3.45}$$

It can be verified that after the distortion of the TX IQ imbalance, the transmitted signal equals the original signal $\overline{\mathbf{s}}$.

The second step is to compensate the RX IQ imbalance, i.e., to calculate

$$\overline{\mathbf{y}} = \frac{\overline{\mathbf{r}} - \xi_r \overline{\mathbf{r}}^*}{1 - |\xi_r|^2} = \overline{\Lambda}\,[\overline{\mathbf{S}}\,\overline{\mathbf{h}} + \overline{\mathbf{w}}], \tag{3.46}$$

where $\overline{\mathbf{w}} = [w(0), w(1), \cdots, w(N-1)]^T$.

The last step is to compensate the frequency offset, i.e., to get

$$\hat{\overline{\mathbf{y}}} = \overline{\Lambda}^{-1}\,\overline{\mathbf{y}} = \overline{\mathbf{S}}\,\overline{\mathbf{h}} + \overline{\mathbf{w}}. \tag{3.47}$$

After that, conventional detection method can be used to recover $\overline{\mathbf{s}}$.

Another way to compensate the frequency offset and TX and RX IQ imbalances is to use adaptive filters [22]. In the following, we describe this adaptive filtering method.

Define the vector $\overline{\mathbf{r}}^{(1)}$ as

$$\begin{aligned} \overline{\mathbf{r}}^{(1)} &\triangleq \overline{\Lambda}^{-1}\,\overline{\mathbf{r}} = \overline{\mathbf{p}} + \xi_r \underbrace{\overline{\Lambda}^{-2}\overline{\mathbf{p}}^*}_{\triangleq \overline{\mathbf{q}}} + \underbrace{\overline{\mathbf{w}} + \xi_r \overline{\Lambda}^{-2}\overline{\mathbf{w}}^*}_{\triangleq \overline{\mathbf{v}}^{(1)}} \\ &= \overline{\mathbf{p}} + \xi_r \overline{\mathbf{q}} + \overline{\mathbf{v}}^{(1)}, \end{aligned} \tag{3.48}$$

and $\mathbf{r}^{(2)}$ as

$$\begin{aligned} \overline{\mathbf{r}}^{(2)} &\triangleq \overline{\Lambda}^{-1}\,\overline{\mathbf{r}}^* = \xi_r^* \overline{\mathbf{p}} + \underbrace{\overline{\Lambda}^{-2}\overline{\mathbf{p}}^*}_{=\overline{\mathbf{q}}} + \underbrace{\overline{\Lambda}^{-2}\overline{\mathbf{w}}^* + \xi_r^* \overline{\mathbf{w}}}_{\triangleq \overline{\mathbf{v}}^{(2)}} \\ &= \xi_r^* \overline{\mathbf{p}} + \overline{\mathbf{q}} + \overline{\mathbf{v}}^{(2)}. \end{aligned} \tag{3.49}$$

Transform the vectors $\overline{\mathbf{r}}^{(1)}$, $\overline{\mathbf{r}}^{(2)}$, $\overline{\mathbf{p}}$, $\overline{\mathbf{q}}$, $\overline{\mathbf{v}}^{(1)}$, and $\overline{\mathbf{v}}^{(2)}$ to the frequency domain, i.e., do $\mathbf{r}^{(1)} = \mathbf{F}\overline{\mathbf{r}}^{(1)}$, $\mathbf{r}^{(2)} = \mathbf{F}\overline{\mathbf{r}}^{(2)}$, $\mathbf{p} = \mathbf{F}\overline{\mathbf{p}}$, $\mathbf{q} = \mathbf{F}\overline{\mathbf{q}}$, $\mathbf{v}^{(1)} = \mathbf{F}\overline{\mathbf{v}}^{(1)}$, and $\mathbf{v}^{(2)} = \mathbf{F}\overline{\mathbf{v}}^{(2)}$, where $\mathbf{F}$ is a size N DFT matrix. Multiplying both sides of (3.48) and (3.49) by a DFT matrix $\mathbf{F}$ and writing them in a vector form, we have

$$\begin{bmatrix} \mathbf{r}^{(1)} \\ \mathbf{r}^{(2)} \end{bmatrix} = \begin{bmatrix} 1 & \xi_r \\ \xi_r^* & 1 \end{bmatrix} \begin{bmatrix} \mathbf{p} \\ \mathbf{q} \end{bmatrix} + \begin{bmatrix} \mathbf{v}^{(1)} \\ \mathbf{v}^{(2)} \end{bmatrix}. \tag{3.50}$$

The k-th components of $\mathbf{r}^{(1)}$ and $\mathbf{r}^{(2)}$, i.e., $R_k^{(1)}$ and $R_k^{(2)}$, can be written as

$$\begin{bmatrix} R_k^{(1)} \\ R_k^{(2)} \end{bmatrix} = \begin{bmatrix} 1 & \xi_r \\ \xi_r^* & 1 \end{bmatrix} \begin{bmatrix} P_k \\ Q_k \end{bmatrix} + \begin{bmatrix} V_k^{(1)} \\ V_k^{(2)} \end{bmatrix}, \tag{3.51}$$

where P_k and Q_k are the k-th component of $\mathbf{p}$ and $\mathbf{q}$, respectively, and $V_k^{(1)}$ and $V_k^{(2)}$ are the k-th component of $\mathbf{v}^{(1)}$ and $\mathbf{v}^{(2)}$, respectively.

Since in (3.44), $\overline{\mathbf{p}}$ is defined as $\overline{\mathbf{p}} = \overline{\mathbf{S}}\,\overline{\mathbf{h}} + \xi_t \overline{\mathbf{S}}^* \overline{\mathbf{h}}$, we have that $P_k = H_k(S_k + \xi_t S_{N-k}^*)$, where H_k is the channel coefficient in the k-th subcarrier, S_k is the k-th component of the DFT of $\bar{\mathbf{s}}$. Substitute it into (3.51), we have

$$\begin{bmatrix} R_k^{(1)} \\ R_k^{(2)} \end{bmatrix} = \begin{bmatrix} H_k & \xi_t H_k & \xi_r \\ \xi_r^* H_k & \xi_r^* \xi_t H_k & 1 \end{bmatrix} \begin{bmatrix} S_k \\ S_{N-k}^* \\ Q_k \end{bmatrix} + \begin{bmatrix} V_k^{(1)} \\ V_k^{(2)} \end{bmatrix}. \tag{3.52}$$

Similarly, we can write $(R_{N-k}^{(1)})^*$ and $(R_{N-k}^{(2)})^*$ as

$$\begin{bmatrix} (R_{N-k}^{(1)})^* \\ (R_{N-k}^{(2)})^* \end{bmatrix} = \begin{bmatrix} H_{N-k}^* & \xi_t^* H_{N-k}^* & \xi_r^* \\ \xi_r H_{N-k}^* & \xi_r \xi_t^* H_{N-k}^* & 1 \end{bmatrix} \begin{bmatrix} S_{N-k}^* \\ S_k \\ Q_{N-k}^* \end{bmatrix} + \begin{bmatrix} (V_{N-k}^{(1)})^* \\ (V_{N-k}^{(2)})^* \end{bmatrix}. \tag{3.53}$$

Combine (3.52) and (3.53), we have

$$\underbrace{\begin{bmatrix} R_k^{(1)} \\ R_k^{(2)} \\ (R_{N-k}^{(1)})^* \\ (R_{N-k}^{(2)})^* \end{bmatrix}}_{\triangleq \mathbf{r}_{k,N-k}} = \underbrace{\begin{bmatrix} H_k & \xi_t H_k & \xi_r & 0 \\ \xi_r^* H_k & \xi_r^* \xi_t H_k & 1 & 0 \\ \xi_t^* H_{N-k}^* & H_{N-k}^* & 0 & \xi_r^* \\ \xi_r \xi_t^* H_{N-k}^* & \xi_r H_{N-k}^* & 0 & 1 \end{bmatrix}}_{\triangleq \mathbf{H}_{k,N-k}} \underbrace{\begin{bmatrix} S_k \\ S_{N-k}^* \\ Q_k \\ Q_{N-k}^* \end{bmatrix}}_{\triangleq \mathbf{s}_{k,N-k}} + \begin{bmatrix} V_k^{(1)} \\ V_k^{(2)} \\ (V_{N-k}^{(1)})^* \\ (V_{N-k}^{(2)})^* \end{bmatrix}. \tag{3.54}$$

As a result, to get S_k and S_{N-k}^*, we can compensate $\mathbf{r}_{k,N-k}$ as

$$\tilde{\mathbf{s}}_{k,N-k} = \mathbf{H}_{k,N-k}^{-1} \mathbf{r}_{k,N-k}, \tag{3.55}$$

and S_k and S_{N-k}^* are just the first two components of $\tilde{\mathbf{s}}_{k,N-k}$.

If the channel and TX and RX IQ impairments are unknown, adaptive filtering and training symbols can be used to recover the transmitted symbols from $\mathbf{r}_{k,N-k}$ [22]. Assuming that M_{tr} training symbols are used, for the m-th symbol (or the m-th iteration), the filtered symbol can be written as

$$\tilde{S}_{k,m} = w_{k,m}^{(1)} R_k^{(1)} + w_{k,m}^{(2)} R_k^{(2)} + w_{k,m}^{(3)} (R_{N-k}^{(1)})^* + w_{k,m}^{(4)} (R_{N-k}^{(2)})^* \tag{3.56}$$

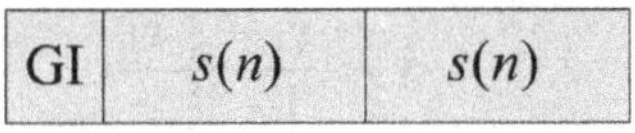

Fig. 3.2 The two repeated sequences for frequency offset and IQ estimation

where $w_{k,m}^{(1)}$, $w_{k,m}^{(2)}$, $w_{k,m}^{(3)}$, and $w_{k,m}^{(4)}$ are the four weights for the m-th iteration for the transmitted symbol in the k-th subcarrier, and they are updated as

$$w_{k,m+1}^{(1)} = w_{k,m}^{(1)} + \mu_{LMS} e_{k,m} R_k^{(1)} \tag{3.57}$$

$$w_{k,m+1}^{(2)} = w_{k,m}^{(2)} + \mu_{LMS} e_{k,m} R_k^{(2)} \tag{3.58}$$

$$w_{k,m+1}^{(3)} = w_{k,m}^{(3)} + \mu_{LMS} e_{k,m} (R_{N-k}^{(1)})^* \tag{3.59}$$

$$w_{k,m+1}^{(4)} = w_{k,m}^{(4)} + \mu_{LMS} e_{k,m} (R_{N-k}^{(2)})^*, \tag{3.60}$$

where μ_{LMS} is the step size and $e_{k,m} = S_{k,m} - \tilde{S}_{k,m}$.

After training, $w_{k,m}^{(1)}$, $w_{k,m}^{(2)}$, $w_{k,m}^{(3)}$, and $w_{k,m}^{(4)}$ can be used to estimate the transmitted symbol when data symbols are sent. Please be noticed, that in order to use this adaptive filtering method, the frequency offset $\Delta\omega$ needs to be known, however, the TX and RX IQ imbalances ξ_t and ξ_r can be unknown.

3.4.2 Parameter Estimation

In this subsection, we discuss how to estimate the frequency offset $\Delta\omega$, and the TX and RX IQ imbalances ξ_t and ξ_r. Chung and Phoong[3] proposed a method based on the definition of CRE [19] for MIMO system, here for clarity, we tailor it to single transmit and receive antenna systems.

Assume that two repeated training sequences like that shown in Fig. 3.2 are available. Based on (3.44), if we denote the training sequence as $\overline{\mathbf{s}}$, and denote

$$\overline{\mathbf{X}} = \overline{\mathbf{S}} + \xi_t \overline{\mathbf{S}}^* \tag{3.61}$$

and

$$\overline{\mathbf{y}} = \overline{\Lambda}\,\overline{\mathbf{X}}\,\overline{\mathbf{h}}, \tag{3.62}$$

then the received signal during the two repeated sequences are

$$\overline{\mathbf{r}}_1 = \overline{\mathbf{y}} + \xi_r \overline{\mathbf{y}}^* + \overline{\mathbf{v}}_1 \tag{3.63}$$

$$\overline{\mathbf{r}}_2 = e^{j\Delta\omega N}\overline{\mathbf{y}} + \xi_r e^{-j\Delta\omega N}\overline{\mathbf{y}}^* + \overline{\mathbf{v}}_2 \tag{3.64}$$

In case ξ_r is perfectly known, then after the RX IQ imbalance compensation, we have

$$\tilde{\mathbf{y}} = \frac{\overline{\mathbf{r}}_1 - \xi_r \overline{\mathbf{r}}_1}{1 - |\xi_r|^2} \tag{3.65}$$

$$e^{j\Delta\omega N}\tilde{\mathbf{y}} = \frac{\overline{\mathbf{r}}_2 - \xi_r \overline{\mathbf{r}}_2}{1 - |\xi_r|^2}. \tag{3.66}$$

The frequency offset $\Delta\omega$ can be estimated as

$$\Delta\tilde{\omega} = \frac{\measuredangle\{[\overline{\mathbf{r}}_1 - \xi_r \overline{\mathbf{r}}_1]^H [\overline{\mathbf{r}}_2 - \xi_r \overline{\mathbf{r}}_2]\}}{N}. \tag{3.67}$$

Assume that $\Delta\omega$, ξ_t, and ξ_r are perfectly known, based on (3.62), the channel can be estimated as

$$\tilde{\overline{\mathbf{h}}} = \frac{1}{2}\overline{\mathbf{X}}^{-1}\overline{\Lambda}^{-1}\left(\frac{\overline{\mathbf{r}}_1 - \xi_r \overline{\mathbf{r}}_1}{1 - |\xi_r|^2} + e^{-j\Delta\omega N} \cdot \frac{\overline{\mathbf{r}}_2 - \xi_r \overline{\mathbf{r}}_2}{1 - |\xi_r|^2}\right). \tag{3.68}$$

Having $\tilde{\overline{\mathbf{h}}}$, we can define CRE in the same way as (3.37). If we write $\hat{\overline{\mathbf{r}}}$ as

$$\hat{\overline{\mathbf{r}}} = \overline{\Lambda}^{-1} (\overline{\mathbf{r}}_1 + e^{-j\Delta\omega N} \overline{\mathbf{r}}_2) \tag{3.69}$$

and by replacing $\overline{\mathbf{r}}$ with $\hat{\overline{\mathbf{r}}}$, we can write the CRE in the same format as (3.38). Then, the TX and RX IQ imbalance can be estimated using the same method as in Eq. (3.41).

To start with, we can assume that $\xi_r \ll 1$, and estimate the frequency offset as

$$\Delta\tilde{\omega} = \frac{\measuredangle\{\overline{\mathbf{r}}_1^H \overline{\mathbf{r}}_2\}}{N}. \tag{3.70}$$

Substituting the estimated frequency offset got in (3.70) into the CRE, we can get initial estimate of the TX and RX IQ imbalances $\tilde{\xi}_t$ and $\tilde{\xi}_r$. Having the initial estimated, we can use them to enhance the frequency offset estimation.

Since the TX and RX IQ imbalances (ξ_t and ξ_r) and the frequency offset ($\Delta\omega$) are intervened. The complexity of joint estimation of these three parameters is very high, [13] proposed an EM (Expectation Maximization) method to estimate these three method. Please refer to [13] for the details.

3.5 Joint TX and RX IQ Imbalance Estimation and Compensation for Multiple Antenna Systems

For the spatial multiplexing system, as shown in Sect. 2.5, if IQ imbalance exists, due to the interference between the symmetric subcarriers, the original $N_t \times N_r$ system now becomes a $2N_t \times 2N_r$ system, with expanded channel matrix including the

IQ imbalance. Specifically, because the IQ imbalance is now frequency independent, we denote $\xi_{t,m}$ (or $\xi_{r,n}$) as the IQ imbalance for the m-th TX (or n-th RX) antenna,[1] then, according to Eq. (2.69), the received signal is now equal to

$$\underbrace{\begin{bmatrix} \mathbf{r}_k \\ \mathbf{r}^*_{N-k} \end{bmatrix}}_{\triangleq \mathbf{r}_{r,N-k}} = \underbrace{\begin{bmatrix} \mathbf{H}_k + \Xi_r \mathbf{H}^*_{N-k} \Xi^*_t & \mathbf{H}_k \Xi_t + \Xi_r \mathbf{H}^*_{N-k} \\ \Xi^*_r \mathbf{H}_k + \mathbf{H}^*_{N-k} \Xi^*_t & \Xi^*_r \mathbf{H}_k \Xi_t + \mathbf{H}^*_{N-k} \end{bmatrix}}_{\triangleq \mathbf{H}_{k,N-k}} \underbrace{\begin{bmatrix} \mathbf{s}_k \\ \mathbf{s}^*_{N-k} \end{bmatrix}}_{\triangleq \mathbf{s}_{k,N-k}} + \mathbf{v}_{k,N-k}, \quad (3.71)$$

where

$$\begin{aligned} \mathbf{r}_k &= [R_{1,k}, R_{2,k}, \cdots, R_{N_r,k}]^T \\ \mathbf{s}_k &= [S_{1,k}, S_{2,k}, \cdots, S_{N_t,k}]^T \\ \Xi_t &= \mathrm{diag}\{\xi_{t,1}, \xi_{t,2}, \cdots, \xi_{t,N_t}\} \\ \Xi_r &= \mathrm{diag}\{\xi_{r,1}, \xi_{r,2}, \cdots, \xi_{r,N_r}\} \end{aligned}$$

and $\mathbf{H}_k$ is an $N_r \times N_t$ matrix whose i-th row and j-th column is the channel coefficient between the i-th received and j-th transmit antenna at subcarrier k. $\mathbf{v}_{k,N-k}$ is the colored noise equals

$$\mathbf{v}_{k,N-k} = \begin{bmatrix} \mathbf{I}_{N_r} & \Xi_r \\ \Xi^*_r & \mathbf{I}_{N_r} \end{bmatrix} \begin{bmatrix} \mathbf{w}_k \\ \mathbf{w}^*_{N-k} \end{bmatrix}. \quad (3.72)$$

For the compensation, if the IQ imbalances for the transmit and receive antennas, i.e., Ξ_t and Ξ_r, are known, the compensation can be carried out in the same way as that for single antenna systems. If they are perfectly compensated, the $2N_t \times 2N_r$ system due to IQ imbalance can now be approximated as the original $N_t \times N_r$ system, and conventional MIMO detection method can be applied.

If Ξ_t and Ξ_r are not known, the same method as described in Sect. 3.3 can be used to estimate the expanded channel matrix $\mathbf{H}_{k,N-k}$ defined in (3.71). With the estimated channel matrix, linear detection method can be used to recover the transmitted signal.

For spatial multiplexing MIMO with IQ imbalance, please see for example [3, 5, 24, 29, 30] and the references therein.

For the space-time coded system with IQ imbalances at both TX and RX sides, the received signal can be written as Eq. (2.79), which equals

$$\begin{aligned} r_1 &= as_1 + bs_2 + cs_1^* + ds_2^* + v_1 \\ r_2 &= ds_1 - cs_2 + bs_1^* - as_2^* + v_2 \end{aligned} \quad (3.73)$$

[1]With a slight abuse of notations, here we use $\xi_{t,m}$ to denote the IQ imbalance for the m-th TX antennas, instead of the IQ imbalance at the m-th subcarrier, as is used in previous sections.

where a,b,c, and d are defined in (2.80)–(2.83), respectively. In [18], to estimate the parameters a, b, c, and d, the authors proposed a training scheme using two space-time codes. The symbols transmitted in the first space-time code are

$$s_1 = s_p,\ s_2 = s_p^*; \tag{3.74}$$

and the corresponding received signals in two time slots are r_1 and r_2. The symbols transmitted in the second space-time codes are

$$s_3 = s_p,\ s_4 = s_p, \tag{3.75}$$

and the corresponding received symbols are r_3 and r_4.

Based on (3.73), we have that

$$\underbrace{\begin{bmatrix} r_1 \\ r_2 \\ r_3 \\ r_4 \end{bmatrix}}_{\triangleq \mathbf{r}_p} = \underbrace{\begin{bmatrix} s_p & s_p^* & s_p^* & s_p \\ -s_p & s_p^* & -s_p^* & s_p \\ s_p & s_p & s_p^* & s_p^* \\ -s_p^* & s_p^* & -s_p & s_p \end{bmatrix}}_{\triangleq \mathbf{S}_p} \underbrace{\begin{bmatrix} a \\ b \\ c \\ d \end{bmatrix}}_{\triangleq \Theta} + \underbrace{\begin{bmatrix} v_1 \\ v_2 \\ v_3 \\ v_4 \end{bmatrix}}_{\triangleq \mathbf{v}_p}. \tag{3.76}$$

Clearly, if $s_p \neq s_p^*$, the matrix $\mathbf{S}_p$ is full rank, and the parameter vector $\tilde{\Theta} = [\tilde{a}, \tilde{b}, \tilde{c}, \tilde{d}]^T$ can be estimated as

$$\tilde{\Theta} = \mathbf{S}_p^{-1}\mathbf{r}_p. \tag{3.77}$$

Having $\tilde{\Theta}$, when transmitting information symbol, [18] proposed to use the following equation

$$\underbrace{\begin{bmatrix} r_1 \\ r_1^* \\ r_2 \\ r_2^* \end{bmatrix}}_{\triangleq \mathbf{r}} = \underbrace{\begin{bmatrix} \tilde{a} & \tilde{c} & \tilde{b} & \tilde{d} \\ \tilde{c}^* & \tilde{a}^* & \tilde{d}^* & \tilde{b}^* \\ \tilde{d} & \tilde{b} & -\tilde{c} & -\tilde{a} \\ \tilde{b}^* & \tilde{d}^* & -\tilde{a}^* & -\tilde{c}^* \end{bmatrix}}_{\triangleq \mathbf{H}} \underbrace{\begin{bmatrix} s_1 \\ s_1^* \\ s_2 \\ s_2^* \end{bmatrix}}_{\triangleq \mathbf{s}} + \underbrace{\begin{bmatrix} v_1 \\ v_1^* \\ v_2 \\ v_2^* \end{bmatrix}}_{\triangleq \mathbf{v}}, \tag{3.78}$$

and estimated $\mathbf{s}$ as

$$\tilde{\mathbf{s}} = \mathbf{H}^{-1}\mathbf{r}. \tag{3.79}$$

Since the information symbol s_1 and s_2 are the first and third elements in $\mathbf{s}$, full matrix inverse of $\mathbf{H}$ may not be needed.

Another way to deal with the space-time code with IQ imbalance is to decompose (3.73) into real and imaginary parts [18], which equals

$$\begin{bmatrix} r_1^I \\ r_1^Q \\ r_2^I \\ r_2^Q \end{bmatrix} = \begin{bmatrix} a^I + c^I & -a^Q + c^Q & b^I + d^I & -b^Q + d^Q \\ a^Q + c^Q & a^I - c^I & b^Q + d^I & b^I - d^Q \\ d^I + b^I & -d^Q + b^Q & -c^I - a^I & c^Q - a^Q \\ d^Q + b^Q & d^I - b^I & -c^Q - a^Q & -c^I + a^I \end{bmatrix} \underbrace{\begin{bmatrix} s_1^I \\ s_1^Q \\ s_2^I \\ s_2^Q \end{bmatrix}}_{\triangleq \mathbf{s}_{iq}} + \begin{bmatrix} v_1^I \\ v_1^Q \\ v_2^I \\ v_2^Q \end{bmatrix}. \tag{3.80}$$

The advantage of writing the received signal in this way is that for most modulation schemes, such as QAM and some PSK, the elements in $\mathbf{s}_{iq}$ are independent. Then, some blind method can be used to recover the transmitted symbols. For details, please refer to [18].

3.6 Conclusions

Frequency independent IQ imbalance is a reasonable assumption for narrow band systems. In this chapter, we discussed the estimation and compensation of IQ imbalance which is frequency independent. We first looked at the RX side IQ imbalance estimation and compensation, and then looked at the IQ imbalance estimation and compensation at both TX and RX. After that, we discussed the IQ imbalance estimation and compensation together with frequency offset. At the end, the estimation and compensation of IQ imbalance in multiple antenna systems were discussed.

References

1. Y. Yoshida, K. Hayashi, H. Sakai, and W. Bocquet, "Analysis and compensation of transmitter IQ imbalances in OFDMA and SC-FDMA systems," *IEEE Trans. Signal Process.*, vol. 57, no. 8, pp. 3119–3129, Aug. 2009.
2. L. Brotje, S. Vogeler, and K.-D. Kammeyer, "Estimation and correction of transmitter-caused I/Q imbalance in OFDM systems," *Proc. 7th Intl. OFDM Workshop*, pp. 178–182, Sept. 2002.
3. Y.-H. Chung and S.-M. Phoong, "Joint estimation of I/Q imbalance, CFO and channel response for MIMO OFDM systems," *IEEE Trans. Commun.* vol. 58, no. 5, pp. 1485–1492, May 2010.
4. J. Qi and S. Aïssa, "Analysis and compensation of I/Q imbalance in MIMO transmit-receive diversity systems," *IEEE Trans. Commun.*, vol. 58, no. 5, pp. 1546–1556, May 2010.
5. A. Tarighat and A. H. Sayed, "MIMO OFDM receivers for systems with IQ imbalances," *IEEE Trans. Signal Process.*, vol. 53, no. 9, pp. 3583–3596, Sept. 2005.
6. A. Tarighat, R. Bagheri, and A. H. Sayed, "Compensation schemes and performance analysis of IQ imbalances in OFDM receivers," *IEEE Trans. Signal Process.*, vol. 53, no. 8, pp. 3257–3268, Aug. 2005.
7. K.-Y. Sun and C.-C. Chao, "Estimation and compensation of I/Q imbalance in OFDM direct-conversion receivers," *IEEE J. Sel. Topics in Signal Process.*, vol. 3, no. 3, pp. 438–453, Jun. 2009.

8. M. Valkama, M. Renfors, and V. Koivunen, "Advanced methods for I/Q imbalance compensation in communication receivers," *IEEE Trans. Signal Process.*, vol. 49, no. 10. pp. 2335–2344, Oct. 2001.
9. Y. Egashira, Y. Tanabe, and K. Sato, "A novel IQ imbalance compesation method with pilot signals for OFDM system," *Proc. IEEE VTC-Fall*, pp.1–5, 2006.
10. I.-H. Sohn, E.-R. Jeong, and Y. H. Lee, "Data-aided approach to I/Q mismatch and DC offsect compensation in communication receivers," *IEEE Commun. Lett.*, vol. 6, no. 12, pp. 547–549, Dec. 2002.
11. L. Giugno, V. Lottici, and M. Luise, "Efficient compensation of I/Q phase imbalance for digital receivers," *Proc. IEEE ICC*, 2005.
12. W. Namgoong and P. Rabiei, "CLRB-archieving I/Q mismatch estimator for low-IF receiver using repetitive training sequence in the presence of CFO," *IEEE Trans. Commun.*, vol. 60, no. 3, pp. 706–713, Mar. 2012.
13. F. Horlin, A. Bourdoux, and L. V. der Perre, "Low-complexity EM-based joint acquisition of the carrier frquency offset and IQ imbalance," *IEEE Trans. Wireless Commun.*, vol. 7, no. 6, pp. 2212–2220, Jun. 2008.
14. G.-T. Gil, "Nondata-aided I/Q mismatch and DC offset compensation for direct-conversion receivers," *IEEE Trans. Signal Process.*, vol. 56, no. 7, pp. 2662–2668, Jul. 2008.
15. S. Simoens, M. de Courville, F. Bourzeix, and P. de Champs, "New I/Q imbalance modeling and compensation in OFDM systems with frequency offset," *Proc. IEEE PIMRC 2002.*
16. M. Marey, M. Samir, and O. A. Dobre, "EM-based joint channel estimation and IQ imbalances for OFDM systems," *IEEE Trans. Broadcast.*, vo. 58, no. 1, pp. 106–113, Mar. 2012.
17. P. Rykaczewski, M. Valkama, and M. Renfors, "On the connection of I/Q imbalance and channel equalization in direct-conversion tranceivers," *IEEE Trans. Veh. Technol.*, vol. 57, no. 3, pp. 1630–1636, May 2008.
18. Y. Zou, M. Valkama, and M. Renfors, "Digital compensation of I/Q imbalance effects in space-time coded transmit diversity systems," *IEEE Trans. Signal Process.*, vol. 56, no. 6, pp. 2496–2508, Jun. 2008.
19. Y.-H. Chung and S.-M. Phoon, "Channel estimation in the presence of transmitter and receiver I/Q mismatches for OFDM systems," *IEEE Trans. Wireless Commun.*, vol. 8, no. 9, pp. 4476–4479, Sept. 2009.
20. B. Debaillie, P. V. Wesemael, G. Vandersteen, and J. Craninckx, "Calibration of direct-conversion transceivers," *IEEE J. Sel. Topics in Signal Process.*, vol. 3, no. 3, pp. 488–498, Jun. 2009.
21. S. A. Bassam, S. Boumaiza, and F. M. Ghannouchi, "Block-wise estimation of and compensation for I/Q imbalance in direct-conversion transmitters," *IEEE Trans. Signal Process.*, vol. 57, no. 12. pp. 4970–4973, Dec. 2009.
22. D. Tandur and M. Moonen, "Joint adaptive compensation of transmitter and receiver IQ imbalance under carrier frequency offset in OFDM-based systems," *IEEE Trans. Signal Process.*, vol. 55, no. 11, pp. 5246–5252, Nov. 2007.
23. J. Feigin and D. Brady, "Joint transmitter/receiver I/Q imbalance compensation for direct conversion OFDM in packet-switched multipath environments," *IEEE Trans. Signal Process.*, vol. 57, no. 11. pp. 4588–4593, Nov. 2009.
24. T. Schenk, P. Smulders, and E. Fledderus, "Estimation and compensation of TX and RX IQ imbalance in OFDM-based MIMO systems," *Proc. IEEE Radio and Wireless Symposium*, pp. 215–218, 2006.
25. R. Chrabieh and S. Soliman, "IQ imbalance mitigation via unbiased training sequences," *Proc. IEEE Globecom 2007.*
26. A. Tarighat and A. H. Sayed, "Joint compensation of transmitter and receiver impairments in OFDM systems," *IEEE Trans. Wireless Commun.* vol. 6, no. 1, pp. 240–247, Jan. 2007.
27. C.-J. Hsu and W.-H. Sheen, "Joint calibration of transmitter and receiver impairments in direct-conversion radio architecture," *IEEE Trans. Wireless Commun.*, vol. 11, no. 2, pp. 832–841, Feb. 2012.

28. W. Kirkland and K. Teo, "I/Q distortion correction for OFDM direct conversion receiver," *Electron. Lett.*, vol. 39, pp. 131–133, 2003.
29. H. Minn and D. Munoz, "Pilot designs for channel estimation of MIMO OFDM systems with frequency-depedent I/Q imbalances," *IEEE Trans. Commun.*, vol. 58, no. 8, pp. 2252–2264, Aug. 2010.
30. B. Narasimhan, S. Narayanan, H. Minn, and N. Al-Dhahir, "Reduced-complexity baseband compensation of joint Tx/Rx I/Q imbalance in mobile MIMO-OFDM," *IEEE Trans. Wireless Commun.* vol. 9, no. 5, pp. 1720–1728, May 2010.

Chapter 4
Frequency Dependent IQ Imbalance Estimation and Compensation

Abstract For wideband systems, IQ imbalance values may be frequency selective. To do estimation and compensation, the frequency selectivity cannot be neglected. In this chapter, we look at the frequency dependent IQ imbalance estimation and compensation. We first look at the estimation and compensation of IQ imbalance at the transmitter, then look at joint RX IQ imbalance and frequency offset estimation and compensation at the receiver.

4.1 Introduction

For wideband systems, for example, the multi-giga-hertz bandwidth systems in millimeter-wave band [1], it is difficult to design baseband filters to have consistent performance, especially at the band edge. In this case, in order to achieve good performance, dealing with frequency dependent (FD) IQ imbalance becomes necessary [2]. So far, there are many literatures that discussed the FD IQ imbalance, see for example [3–24] and the references therein.

In this chapter, we discuss the estimation and compensation of FD IQ imbalance. We first look at the FD TX IQ imbalance estimation and compensation, and then look at FD RX IQ imbalance estimation and compensation. To estimate and compensate TX IQ imbalance, an ideal receiver without RX IQ imbalance is needed. The TX IQ imbalance is intervened with the unknown channel coefficients. So, to do TX IQ imbalance estimation and compensation, the channel response must be considered. The RX IQ imbalance is intervened with frequency offset, which always exists in the system. So, to do RX IQ imbalance estimation and compensation, the frequency offset must be considered at the same time.

Y. Li, *In-Phase and Quadrature Imbalance: Modeling, Estimation, and Compensation*, SpringerBriefs in Electrical and Computer Engineering, DOI 10.1007/978-1-4614-8618-3_4,

4.2 TX Only IQ Imbalance Estimation and Compensation

As discussed in Sect. 3.1, for standard compliant systems, the TX IQ imbalance is usually estimated and compensated to meet the EVM requirements specified by the standards. In order to do so, digital down converter (DDC) or an ideal/calibrated IQ demodulator need to be used. As shown in Fig. 4.1, we assume that a DDC is used for the estimation purpose. In the following, we introduce the method proposed in [6] to do TX IQ imbalance estimation and compensation.

In Fig. 4.1, the equivalent IQ imbalance model derived in Sect. 2.2.1 is used. The compensation block in Fig. 4.1 has similar structure as the TX IQ imbalance model. To compensate the TX IQ imbalance, after IQ modulator, the output signal $x(n)$ should be a scale of $s(n)$, i.e.,

$$x(n) = as(n), \tag{4.1}$$

where a is a scalar. To achieve this, assume that, after TX IQ compensation, the output signal equals

$$s'(n) = s(n) + f(n) \otimes s^*(n), \tag{4.2}$$

and after IQ modulator, the output signal equals

$$x(n) = s'(n) + \xi_t(n) \otimes (s'(n))^*. \tag{4.3}$$

Substituting (4.2) into (4.3) gives

$$x(n) = s(n) + f(n) \otimes s^*(n) + \xi_t(n) \otimes s^*(n) + \xi_t(n) \otimes f^*(n) \otimes s(n). \tag{4.4}$$

To meet the requirement of Eq. (4.1), we can choose

$$f(n) = -\xi_t(n), \tag{4.5}$$

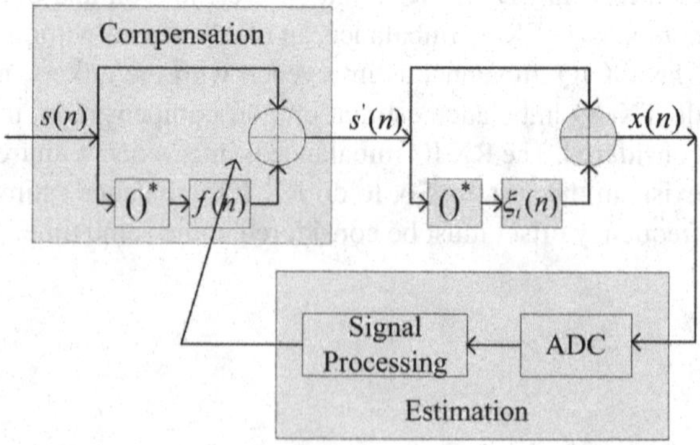

Fig. 4.1 The FD TX IQ estimation and compensation structure

and the scalar a now equals

$$a = 1 - \xi_t(n) \otimes \xi_t^*(n). \tag{4.6}$$

So, if we know the TX IQ imbalance $\xi_t(n)$, using structure shown in Fig. 4.1, we can compensate the TX IQ imbalance.

To estimate the TX IQ imbalance $\xi_t(n)$, the difficulty is that we cannot observe $x(n)$ directly, since the estimation part always introduces unknown channel effects to the observed signal. Assume that the channel is $h(n)$, then observed signal equals

$$y(n) = x(n) \otimes h(n) + w(n), \tag{4.7}$$

where $w(n)$ is the noise.

Assume that the length of the compensation filter $f(n)$ is N_f. At the beginning, we set $f(n) = 0$ for $n = 0, 1, \cdots, N_f - 1$. Then the observed signal equals

$$\begin{aligned} y(n) &= h(n) \otimes s(n) + h(n) \otimes \xi_t(n) \otimes s^*(n) + w(n) \\ &\triangleq h_1(n) \otimes s(n) + h_2(n) \otimes s^*(n) + w(n), \end{aligned} \tag{4.8}$$

where

$$h_1(n) \triangleq h(n) \tag{4.9}$$

$$h_2(n) \triangleq h(n) \otimes \xi_t(n) = h_1(n) \otimes \xi_t(n). \tag{4.10}$$

Also assume that N samples are observed during the estimation, denote them as

$$\mathbf{y}(n) = [y(n), y(n-1), \cdots, y(n-N+1)]^T. \tag{4.11}$$

Then, $\mathbf{y}(n)$ can be written as

$$\mathbf{y}(n) = \mathbf{S}(n)\mathbf{h}_1 + \mathbf{S}^*(n)\mathbf{h}_2 + \mathbf{w}(n), \tag{4.12}$$

where $\mathbf{S}(n)$ is the convolution matrix formed from $s(n)$, $\mathbf{h}_1$ and $\mathbf{h}_2$ are equal to

$$\mathbf{h}_1 = [h_1(0), h_1(1), \cdots, h_1(L_h - 1), \mathbf{0}_{N-L_h}]^T \tag{4.13}$$

$$\mathbf{h}_2 = [h_2(0), h_2(1), \cdots, h_2(L_h - 1), \mathbf{0}_{N-L_h}]^T, \tag{4.14}$$

and $\mathbf{w}(n)$ is the noise vector.

Writing $\mathbf{y}(n)$ in matrix form we have

$$\mathbf{y}(n) = \underbrace{[\mathbf{S}(n)\ \mathbf{S}^*(n)]}_{\triangleq \mathbf{S}_c(n)} \underbrace{\begin{bmatrix} \mathbf{h}_1 \\ \mathbf{h}_2 \end{bmatrix}}_{\triangleq \mathbf{h}_c} + \mathbf{w}(n). \tag{4.15}$$

Using LS method, we have that the estimated $\mathbf{h}_c$ can be written as

$$\tilde{\mathbf{h}}_c = \begin{bmatrix} \tilde{\mathbf{h}}_1 \\ \tilde{\mathbf{h}}_2 \end{bmatrix} = (\mathbf{S}_c(n))^+ \mathbf{y}(n). \tag{4.16}$$

Having $\tilde{\mathbf{h}}_1$ and $\tilde{\mathbf{h}}_2$ and using Eq. (4.10), we get

$$\tilde{\mathbf{h}}_2 = \tilde{\mathbf{H}}_1 \overline{\xi}_t, \tag{4.17}$$

where $\mathbf{H}_1$ is the convolutional matrix of $\tilde{\mathbf{h}}_1$, and

$$\overline{\xi}_t = [\xi_t(0), \xi_t(1), \cdots, \xi_t(L_t - 1), \mathbf{0}_{N-L_t}]^T, \tag{4.18}$$

and L_t is the length of $\xi_t(n)$.

Using LS method again, we get

$$\overline{\xi}_t = \tilde{\mathbf{H}}_1^+ \tilde{\mathbf{h}}_2. \tag{4.19}$$

Having the estimation $\overline{\xi}_t$, based on the relationship between $f(n)$ and $\xi_t(n)$, i.e., Eq. (4.5), the response of the compensation filter can be determined.

Anttila et al. [6] also introduced a blind method to estimate the TX IQ imbalance, where the transmitted signal is assumed to conform the so-called proper property, where the value $\mathbb{E}[s(n)s(n-m)]$ vanishes for all $m \neq 0$. Using this property, we have

$$\mathbf{r}_{sy} = \mathbb{E}[\mathbf{s}(n)y(n)] = \mathbf{R}_s^* \mathbf{h}_1 \tag{4.20}$$

$$\mathbf{r}_{s^*y} = \mathbb{E}[\mathbf{s}^*(n)y(n)] = \mathbf{R}_s \mathbf{h}_2 \tag{4.21}$$

where $\mathbf{R}_s = [\mathbf{s}(n)\mathbf{s}^H(n)]$ and is assumed to be known in advance. Then, the two channels $\mathbf{h}_1$ and $\mathbf{h}_2$ can be estimated as

$$\tilde{\mathbf{h}}_1 = (\mathbf{R}_s^*)^{-1} \mathbf{r}_{sy} \tag{4.22}$$

$$\tilde{\mathbf{h}}_2 = (\mathbf{R}_s)^{-1} \mathbf{r}_{s^*y}. \tag{4.23}$$

Having $\tilde{\mathbf{h}}_1$ and $\tilde{\mathbf{h}}_2$, using the same method as Eq. (4.19), the TX IQ imbalance and corresponding compensation filter response can be estimated.

In [7], using the same principal, adaptive filtering method based on RLS is proposed.

4.3 Joint RX IQ Imbalance and Frequency Offset Estimation and Compensation

As derived in (2.30), when there is FD RX IQ imbalance, the received and transmitted signal has the following relation

$$\begin{aligned} r(n) &= h(n) \otimes s(n) + \xi_r(n) \otimes h^*(n) \otimes s^*(n) \\ &\quad + w(n) + \xi_r(n) \otimes w^*(n). \end{aligned} \tag{4.24}$$

If frequency offset exists, the received and transmitted signal can be written as

$$\begin{aligned} r(n) &= e^{j\Delta\omega n}\left[h(n) \otimes s(n)\right] \\ &\quad + \xi_r(n) \otimes \left(e^{-j\Delta\omega n}\left[h^*(n) \otimes s^*(n)\right]\right) \\ &\quad + e^{j\Delta\omega n} w(n) + \xi_r(n) \otimes \left(e^{-j\Delta\omega n} w^*(n)\right). \end{aligned} \tag{4.25}$$

To compensate the frequency offset and the RX IQ imbalance, we can use the compensation scheme as shown in Fig. 4.2 first proposed in [8]. In this compensation scheme, after the IQ demodulation, in Q branch, there is an L order FIR filter and in I branch there is a $\hat{L} = \frac{L-1}{2}$-sample delay. These are used to compensate the FD IQ imbalance due to the discrepancies in the filters between the I and Q branches in the demodulation. After the filtering and delay, a cross-talk between the I and Q branches is introduced, which is used to compensate the frequency independent (FI) part of the IQ imbalance. After this, the frequency offset is compensated.

In stead of estimating the FD IQ imbalance, the objective now is to estimate the FIR filter coefficient $f(n)$ and the cross-talk coefficient β. To estimate these coefficients, a so-called MPP (Modified Periodic Pilot) is proposed in [8], which is shown in Fig. 4.3, where there are totally M pilot symbols, each has a guard interval (GI) of size L_{gi} and a pilot sequence of size N. The odd-numbered pilot sequence is $s(n)$, while the even-numbered pilot sequence has a $\pi/4$ phase rotate, which is $s(n) \cdot e^{j\pi/4}$.

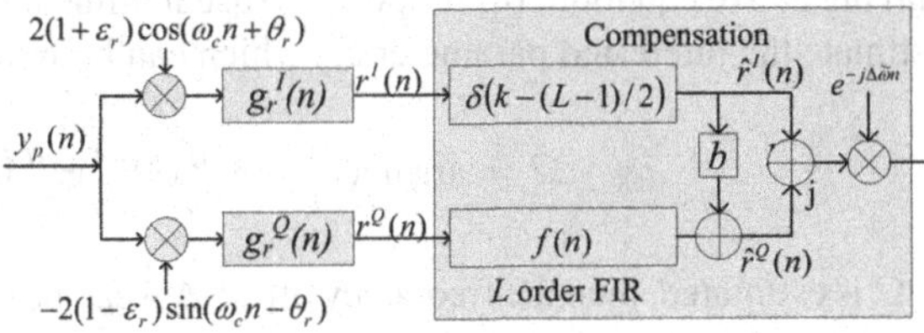

Fig. 4.2 The FD RX IQ compensation structure

Fig. 4.3 The MPP preamble

At the receiver, after removing the GI, we can write the M received sequences as a matrix, which is

$$\overline{\mathbf{R}} = \begin{bmatrix} r_1(0) & r_1(1) & \cdots & r_1(N-1) \\ r_2(0) & r_2(1) & \cdots & r_2(N-1) \\ \vdots & \vdots & \ddots & \vdots \\ r_M(0) & r_M(1) & \cdots & r_M(N-1) \end{bmatrix}, \tag{4.26}$$

where $r_m(n)$ means the n-th received symbol in the m-th pilot sequence after removing GI. Based on (4.25), if the noise $w(n)$ is ignored, when m is an odd number, $r_m(n)$ can be written as

$$r_m(n) = e^{j(m-1)\Omega} \underbrace{\left[e^{j\Delta\omega n}\left(s(n) \otimes h(n)\right)\right]}_{\triangleq \rho(n)}$$
$$+ e^{-j(m-1)\Omega} \underbrace{\left[e^{-j\Delta\omega n}\left(s^*(n) \otimes (\xi_r(n) e^{j\Delta\omega n}) \otimes h^*(n)\right)\right]}_{\triangleq \eta(n)}$$

where $\Omega = \Delta\omega(N + L_{gi})$. When m is an even number, $r_m(n)$ can be written as

$$r_m(n) = e^{j(m-1)\Omega + \pi/4}\rho(n) + e^{-j(m-1)\Omega - \pi/4}\eta(n). \tag{4.27}$$

Then the n-th column of $\overline{\mathbf{R}}$ can be written as

$$\bar{\mathbf{r}}(n) = \begin{bmatrix} r_1(n) \\ r_2(n) \\ \vdots \\ r_M(n) \end{bmatrix} = \underbrace{\begin{bmatrix} 1 & 1 \\ e^{j(\Omega+\pi/4)} & e^{-j(\Omega+\pi/4)} \\ e^{j(2\Omega)} & e^{-j(2\Omega)} \\ \vdots & \vdots \\ e^{j((M-1)\Omega+\pi/4)} & e^{-j((M-1)\Omega+\pi/4)} \end{bmatrix}}_{\triangleq \Theta} \begin{bmatrix} \rho(n) \\ \eta(n) \end{bmatrix}. \tag{4.28}$$

Having above equation, [8] proposed to use nonlinear least square (NLS) method to estimate the interested parameter Ω, which can be written as

$$\tilde{\Omega} = \arg\max_{\Omega} \left[\Theta(\Theta^H \Theta)^{-1} \Theta^H \overline{\mathbf{R}\mathbf{R}}^H\right]. \tag{4.29}$$

If Ω is estimated, then the frequency offset $\Delta\omega$ can be estimated accordingly, denote it as $\Delta\tilde{\omega}$. The phase rotate $\pi/4$ in the even pilot sequences is used to prevent the ill-condition of Θ caused by small frequency offset.

Next step is to estimate the compensation parameters: the L-th order filter coefficients $f(n)$ and the cross-talk coefficient b. We use vector

$$\bar{\mathbf{f}} \triangleq [f(0), f(1), \cdots, f(L-1)]^T$$

to denote the coefficients of the FIR filter in the Q branch. As shown in Fig. 4.2, after compensation, write the signal in the I and Q branch as $\hat{r}^I(n)$ and $\hat{r}^Q(n)$, respectively. Then, based on the pilot symbol structure, $\bar{\mathbf{f}}$ and b should be designed to minimize the following

$$[\bar{\mathbf{f}}\ b]_{opt} = \arg\min_{\bar{\mathbf{f}},b} \left\{ \sum_{m=1}^{M-1} \sum_{n=1}^{N+L-1} \left| \hat{r}_{m+1}(n) - e^{j\Omega_m} \hat{r}_m(n) \right|^2 \right\}, \tag{4.30}$$

where $\hat{r}_m(n) = \hat{r}_m^I(n) + \mathbf{j}\,\hat{r}_m^Q(n)$ is the n-th symbol of the m-th pilot sequence after the compensation, and

$$\Omega_m = \begin{cases} \Omega + \pi/4, & m \text{ is odd} \\ \Omega - \pi/4, & m \text{ is even} \end{cases}. \tag{4.31}$$

Define the size $(N + L - 1) \times L$ matrix $\overline{\mathbf{R}}_m^I$ as

$$\overline{\mathbf{R}}_m^I = \begin{bmatrix} r_m^I(1) & & & \\ r_m^I(2) & r_m^I(1) & & \\ \vdots & r_m^I(2) & \ddots & \\ r_m^I(N) & \vdots & \ddots & r_m^I(1) \\ & r_m^I(N) & \vdots & r_m^I(2) \\ & & \ddots & \vdots \\ & & & r_m^I(N) \end{bmatrix}, \tag{4.32}$$

and define $\overline{\mathbf{R}}_m^Q$ in a similar way. Then after passing the compensation structure, the compensated I and Q vectors

$$\hat{\mathbf{r}}_m^I = [\hat{r}_m^I(1)\ \hat{r}_m^I(2) \cdots \hat{r}_m^I(N + L - 1)]^T$$
$$\hat{\mathbf{r}}_m^Q = [\hat{r}_m^Q(1)\ \hat{r}_m^Q(2) \cdots \hat{r}_m^Q(N + L - 1)]^T$$

equal to

$$\hat{\mathbf{r}}_m^I = \overline{\mathbf{R}}_m^I \mathbf{1}_L \tag{4.33}$$

$$\hat{\mathbf{r}}_m^Q = [\overline{\mathbf{R}}_m^I \mathbf{1}_L\ \ \overline{\mathbf{R}}_m^Q] \begin{bmatrix} b \\ \bar{\mathbf{f}} \end{bmatrix}, \tag{4.34}$$

where $\mathbf{1}_L = [\mathbf{0}_{\frac{L-1}{2}}\ 1\ \mathbf{0}_{\frac{L-1}{2}}]^T$ and $\mathbf{0}_{\frac{L-1}{2}}$ is a size $1 \times (\frac{L-1}{2})$ all zero vector.

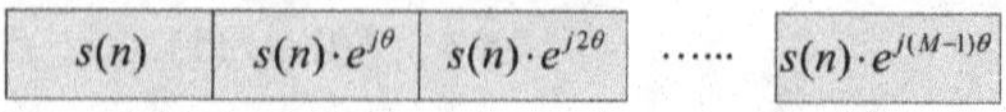

Fig. 4.4 The GPP preamble

Based on the pilot structure, due to the phase rotation, $\hat{\mathbf{r}}^I_{m+1}$ and $\hat{\mathbf{r}}^Q_{m+1}$ can be written as

$$\begin{bmatrix} \hat{\mathbf{r}}^I_{m+1} \\ \hat{\mathbf{r}}^Q_{m+1} \end{bmatrix} = \begin{bmatrix} \cos\Omega_m & -\sin\Omega_m \\ \sin\Omega_m & \cos\Omega_m \end{bmatrix} \begin{bmatrix} \hat{\mathbf{r}}^I_m \\ \hat{\mathbf{r}}^Q_m \end{bmatrix} + \mathbf{n}_m, \tag{4.35}$$

where $\mathbf{n}_m$ is the noise vector. Using (4.33)–(4.35) can be written as

$$\mathbf{A}_m \begin{bmatrix} b \\ \mathbf{f} \end{bmatrix} = \mathbf{B}_m + \mathbf{n}_m, \tag{4.36}$$

where $\mathbf{A}_m$ and $\mathbf{B}_m$ are defined as

$$\mathbf{A}_m = \begin{bmatrix} \sin\Omega_m \overline{\mathbf{R}}^I_m \mathbf{1}_L & \sin\Omega_m \overline{\mathbf{R}}^Q_m \\ \overline{\mathbf{R}}^I_{m+1}\mathbf{1}_L - \cos\Omega_m \overline{\mathbf{R}}^I_m \mathbf{1}_L & \overline{\mathbf{R}}^Q_{m+1} - \cos\Omega_m \overline{\mathbf{R}}^Q_m \end{bmatrix} \tag{4.37}$$

$$\mathbf{B}_m = \begin{bmatrix} \cos\Omega_m \overline{\mathbf{R}}^I_m \mathbf{1}_L - \overline{\mathbf{R}}^I_{m+1}\mathbf{1}_L \\ \sin\Omega_m \overline{\mathbf{R}}^I_m \mathbf{1}_L \end{bmatrix}. \tag{4.38}$$

Write

$$\mathbf{A} = [\mathbf{A}_1\ \mathbf{A}_2\ \cdots \mathbf{A}_{M-1}]^T \tag{4.39}$$

$$\mathbf{B} = [\mathbf{B}_1\ \mathbf{B}_2\ \cdots \mathbf{B}_{M-1}]^T \tag{4.40}$$

then the compensation coefficients can be estimated as [8]

$$\begin{bmatrix} \tilde{b} \\ \tilde{\mathbf{f}} \end{bmatrix} = \mathbf{A}^{\dagger}\mathbf{B}. \tag{4.41}$$

The complexity of the NLS based frequency offset estimation method [8] is high, to lower the complexity, based on the compensation structure of Fig. 4.2, [9] proposed the so-called GPP (Generalized Periodic Pilot) and proposed a linear least square method to estimate the frequency offset. Figure 4.4 shows the GPP Preamble. In the GPP preamble, M pilot sequences are used. Each pilot sequence has N symbols. For each sequence, compared with the sequence that proceeds to it, a phase rotate $e^{j\theta}$ is applied. Due to the delay spread, the first pilot sequence is used as the guard interval. At the receiver, after discarding the guard interval, for $n > N$, if carrier frequency offset, IQ imbalance and noise are not present, we have

$$r(n+N) = e^{j\theta} r(n). \tag{4.42}$$

If frequency offset is counted in, we have

$$r(n+N) = e^{j(\Delta\omega N+\theta)} r(n). \tag{4.43}$$

After the compensation as shown in Fig. 4.2, define two size P received vectors as

$$\hat{\mathbf{r}}_1 = [\hat{r}(K+\hat{L})\ \hat{r}(K+\hat{L}+1)\ \cdots \hat{r}(K+\hat{L}+P-1)]^T, \tag{4.44}$$

$$\hat{\mathbf{r}}_2 = [\hat{r}(K+\hat{L}+N)\ \hat{r}(K+\hat{L}+N+1)\ \cdots \hat{r}(K+\hat{L}+N+P-1)]^T, \tag{4.45}$$

where $K > N$, $\hat{L} = \frac{L-1}{2}$ and L is the length of the FIR filter in the Q branch as shown in Fig. 4.2.

Based on relation in (4.43), the compensation coefficients and frequency offset estimation can be jointly optimized to minimize the following

$$[\Delta\omega, b, \bar{\mathbf{f}}]_{opt} = \arg\min_{\Delta\omega, b, \bar{\mathbf{f}}} \|\hat{\mathbf{r}}_2 - e^{j(\Delta\omega N+\theta)}\hat{\mathbf{r}}_1\|^2. \tag{4.46}$$

According to the compensation structure, $\hat{\mathbf{r}}_1$ and $\hat{\mathbf{r}}_2$ can be written as

$$\hat{\mathbf{r}}_1 = \bar{\mathbf{r}}_1^I + \mathbf{j}\,(\overline{\mathbf{R}}_1^Q \bar{\mathbf{f}} + b\bar{\mathbf{r}}_1^I) \tag{4.47}$$

$$\hat{\mathbf{r}}_2 = \bar{\mathbf{r}}_2^I + \mathbf{j}\,(\overline{\mathbf{R}}_2^Q \bar{\mathbf{f}} + b\bar{\mathbf{r}}_2^I), \tag{4.48}$$

where $\bar{\mathbf{r}}_1^I$ and $\bar{\mathbf{r}}_2^I$ are defined as

$$\bar{\mathbf{r}}_1^I = [r^I(K)\ r^I(K+1)\cdots r^I(K+P-1)]^T \tag{4.49}$$

$$\bar{\mathbf{r}}_2^I = [r^I(K+N)\ r^I(K+N+1)\cdots r^I(K+N+P-1)]^T. \tag{4.50}$$

$\overline{\mathbf{R}}_1^Q$ and $\overline{\mathbf{R}}_2^Q$ are $P \times L$ matrices, whose elements in the p-th row and q-th column are equal to $r^Q(K+\hat{L}+p-q)$ and $r^Q(K+N+\hat{L}+p-q)$, respectively.

Substituting (4.47) and (4.48) into (4.46), we have

$$\begin{bmatrix}\bar{\mathbf{r}}_1^I & -\overline{\mathbf{R}}_1^Q\end{bmatrix}\begin{bmatrix}\cos(\Delta\omega N+\theta) - b\sin(\Delta\omega N+\theta)\\ \bar{\mathbf{f}}\sin(\Delta\omega N+\theta)\end{bmatrix} = \bar{\mathbf{r}}_2^I, \tag{4.51}$$

$$\begin{bmatrix}\bar{\mathbf{r}}_2^I & \overline{\mathbf{R}}_2^Q\end{bmatrix}\begin{bmatrix}\cos(\Delta\omega N+\theta) + b\sin(\Delta\omega N+\theta)\\ \bar{\mathbf{f}}\sin(\Delta\omega N+\theta)\end{bmatrix} = \bar{\mathbf{r}}_1^I. \tag{4.52}$$

Combining above two equations, we have

$$\underbrace{\begin{bmatrix} \bar{\mathbf{r}}_1^I & \mathbf{0} & -\bar{\mathbf{R}}_1^Q \\ \mathbf{0} & \bar{\mathbf{r}}_2^I & \bar{\mathbf{R}}_2^Q \end{bmatrix}}_{\triangleq \mathbf{A}} \underbrace{\begin{bmatrix} \cos(\Delta\omega N + \theta) - b\sin(\Delta\omega N + \theta) \\ \cos(\Delta\omega N + \theta) + b\sin(\Delta\omega N + \theta) \\ \bar{\mathbf{f}}\sin(\Delta\omega N + \theta) \end{bmatrix}}_{\triangleq \mathbf{u}} = \underbrace{\begin{bmatrix} \bar{\mathbf{r}}_2^I \\ \bar{\mathbf{r}}_1^I \end{bmatrix}}_{\triangleq \bar{\mathbf{r}}^I}. \tag{4.53}$$

Then, the vector $\mathbf{u}$ can be estimated as [9]

$$\mathbf{u} = \mathbf{A}^{\dagger}\bar{\mathbf{r}}^I. \tag{4.54}$$

Having $\mathbf{u}$, the frequency offset can be calculated as

$$\Delta\tilde{\omega} = \frac{1}{N}\left\{\arccos\left(\frac{u(1)+u(2)}{2}\right) - \theta\right\}, \tag{4.55}$$

and the estimates of b and $\bar{\mathbf{f}}$ are

$$\tilde{b} = \frac{u(2)-u(1)}{2\sin(\Delta\tilde{\omega}N + \theta)}, \tag{4.56}$$

$$\tilde{\bar{\mathbf{f}}} = \frac{1}{\sin(\Delta\tilde{\omega}N + \theta)}\left[u(3)\; u(4) \cdots u(L+2)\right], \tag{4.57}$$

respectively.

Based on the GPP structure, a better method is proposed in [10] to estimate the frequency offset.

In [10], starting from Eq. (4.25), the received signal $r(n)$ can be written as

$$r(n) = e^{j\Delta\omega n}\left[s(n) \otimes h(n)\right] + e^{-j\Delta\omega n}[s^*(n) \otimes \underbrace{(\xi_r(n)e^{j\Delta\omega n}) \otimes h^*(n)}_{\triangleq \eta(n)}] + v(n), \tag{4.58}$$

where $v(n) = e^{j\Delta\omega n}w(n) + \xi_r(n) \otimes (e^{-j\Delta\omega n}w^*(n))$.

For $N \le n \le (M-2)N$, based on (4.58), we have

$$\underbrace{\begin{bmatrix} r(n) \\ r(n+N) \\ r(n+2N) \end{bmatrix}}_{\triangleq \mathbf{r}(n)} = \underbrace{\begin{bmatrix} 1 & 1 \\ e^{j(\Delta\omega N+\theta)} & e^{-j(\Delta\omega N+\theta)} \\ e^{j(2\Delta\omega N+\theta)} & e^{-j(2\Delta\omega N+\theta)} \end{bmatrix}}_{\triangleq \mathbf{A}(\Delta\omega)} \underbrace{\begin{bmatrix} e^{j\Delta\omega n}[s(n) \otimes h(n)] \\ e^{-j\Delta\omega n}[s^*(n) \otimes \eta(n)] \end{bmatrix}}_{\triangleq \mathbf{x}(n)} + \underbrace{\begin{bmatrix} v(n) \\ v(n+N) \\ v(n+2N) \end{bmatrix}}_{\triangleq \mathbf{v}(n)}. \tag{4.59}$$

Define the $3 \times (M-3)N$ matrix $\mathbf{R}$ as

$$\mathbf{R} = [\mathbf{r}(N)\; \mathbf{r}(N+1) \cdots \mathbf{r}((M-2)N-1)], \tag{4.60}$$

then $\mathbf{R}$ can be written as

$$\mathbf{R} = \mathbf{A}(\Delta\omega)\mathbf{X} + \mathbf{V}, \tag{4.61}$$

where $\mathbf{X}$ equals

$$\mathbf{X} = [\mathbf{x}(N)\ \mathbf{x}(N+1)\cdots\mathbf{x}((M-2)N-1)], \tag{4.62}$$

and $\mathbf{V}$ equals

$$\mathbf{V} = [\mathbf{v}(N)\ \mathbf{v}(N+1)\cdots\mathbf{v}((M-2)N-1)]. \tag{4.63}$$

Having $\mathbf{R}$, we can calculate $\mathbf{R}\mathbf{R}^H$, which equals

$$\mathbf{R}\mathbf{R}^H = \mathbf{A}(\Delta\omega)\mathbf{X}\mathbf{X}^H\mathbf{A}^H(\Delta\omega) + \mathbf{V}\mathbf{V}^H. \tag{4.64}$$

Since $\mathbf{A}(\Delta\omega)$ is a 3×2 matrix, and $\mathbf{X}\mathbf{X}^H$ is usually full rank, we can find a 3×1 vector $\mathbf{a} = [a_0\ a_1\ a_2]$, such that

$$\mathbf{b}^H\mathbf{A}(\Delta\omega)\mathbf{X}\mathbf{X}^H\mathbf{A}^H(\Delta\omega)\mathbf{b} = 0, \tag{4.65}$$

which implies

$$\mathbf{A}^H(\Delta\omega)\mathbf{b} = 0. \tag{4.66}$$

Based on the structure of $\mathbf{A}(\Delta\omega)$ defined in (4.59), we have

$$b_0 + b_1 e^{-j(\Delta\omega N+\theta)} + b_2 e^{-2j(\Delta\omega N+\theta)} = 0 \tag{4.67}$$

$$b_0 + b_1 e^{j(\Delta\omega N+\theta)} + b_2 e^{2j(\Delta\omega N+\theta)} = 0. \tag{4.68}$$

Treat b_0 as known, then b_1 and b_2 can be represented by b_0 as

$$b_1 = -2\cos(\Delta\omega N + \theta), \tag{4.69}$$

$$b_2 = b_0, \tag{4.70}$$

respectively.

Considering this constraint, we can define a 3×2 matrix $\mathbf{B}$, and use $\mathbf{B}$ to represent the vector $\mathbf{b}$, which is

$$\mathbf{b} = \underbrace{\begin{bmatrix} \frac{1}{\sqrt{2}} & 0 \\ 0 & 1 \\ \frac{1}{\sqrt{2}} & 0 \end{bmatrix}}_{\triangleq \mathbf{B}} \triangleq \underbrace{\begin{bmatrix} c_0 \\ c_1 \end{bmatrix}}_{\mathbf{c}}. \tag{4.71}$$

Clearly, we can see that $\mathbf{B}^H\mathbf{B} = \mathbf{I}_2$.

Based on the definition of $\mathbf{B}$ and $\mathbf{c}$, $\cos(\Delta\omega N + \theta)$ can be represented as

$$\cos(\Delta\omega N + \theta) = -\frac{c_1}{\sqrt{2}c_0}. \tag{4.72}$$

Substituting it into (4.65), we have

$$\mathbf{c}^H \mathbf{B}^H \mathbf{A}(\Delta\omega)\mathbf{X}\mathbf{X}^H \mathbf{A}^H(\Delta\omega)\mathbf{B}\mathbf{c} = 0. \tag{4.73}$$

By assuming that the elements in the noise vector $\mathbf{v}(n)$ are i.i.d., we have that

$$\mathbf{V}\mathbf{V}^H \approx \sigma^2 \mathbf{I}_3. \tag{4.74}$$

Calculating the 2×2 matrix

$$\begin{aligned}\mathbf{B}^H\mathbf{R}\mathbf{R}^H\mathbf{B} &= \mathbf{B}^H\mathbf{A}(\Delta\omega N + \theta)\mathbf{X}\mathbf{X}^H\mathbf{A}^H(\Delta\omega N + \theta)\mathbf{B} + \sigma^2\mathbf{B}^H\mathbf{B} \\ &= \mathbf{B}^H\mathbf{A}(\Delta\omega N + \theta)\mathbf{X}\mathbf{X}^H\mathbf{A}^H(\Delta\omega N + \theta)\mathbf{B} + \sigma^2\mathbf{I}_2.\end{aligned} \tag{4.75}$$

Because of (4.73) and (4.75), we have that $\mathbf{c}$ should be the eigenvector of $\mathbf{B}^H\mathbf{R}\mathbf{R}^H\mathbf{B}$ corresponding to its smallest eigenvalue. Denote it as $\tilde{\mathbf{c}} = [\tilde{c}_1\ \tilde{c}_2]^T$, then the estimated frequency offset equals

$$\Delta\tilde{\omega} = \frac{1}{N}\left(\arccos\left(\Re\left\{\frac{-\tilde{c}_2}{\sqrt{2}\tilde{c}_1}\right\}\right) - \theta\right). \tag{4.76}$$

Since arccos is in the range of $[0\ \pi)$, the estimated frequency offset is in the range

$$-\frac{\theta}{N} \le \Delta\tilde{\omega} < -\frac{\theta}{N} + \frac{\pi}{N}, \tag{4.77}$$

and this estimation method is called "Cosine estimator" in [10]. To improve the estimation range and accuracy, a "Tangent estimator" is further proposed in [10].

It is shown in [10] that the frequency offset estimation obtained by using this method is better than that in [9]. Having the frequency offset estimation, we can revise the method proposed in [9] to get the compensation filter coefficients. Specifically, we define the matrix $\mathbf{A}$ in this case as

$$\mathbf{A} = \begin{bmatrix} \bar{\mathbf{r}}_1^I\ \overline{\mathbf{R}}_1^Q \\ \bar{\mathbf{r}}_2^I\ \overline{\mathbf{R}}_2^Q \end{bmatrix}, \tag{4.78}$$

where $\bar{\mathbf{r}}_1^I$, $\bar{\mathbf{r}}_2^I$, $\overline{\mathbf{R}}_1^Q$, and $\overline{\mathbf{R}}_2^Q$ are the same as those used in (4.51) and (4.52). Also, we define vector $\bar{\mathbf{r}}^I$ in this case as

$$\bar{\mathbf{r}}^I = \begin{bmatrix} \bar{\mathbf{r}}_1^I \cos(\Delta\tilde{\omega}N + \theta) - \bar{\mathbf{r}}_2^I \\ \bar{\mathbf{r}}_1^I - \bar{\mathbf{r}}_2^I \cos(\Delta\tilde{\omega}N + \theta) \end{bmatrix} \cdot \frac{1}{\sin(\Delta\tilde{\omega}N + \theta)}. \tag{4.79}$$

Then, we calculate the

$$\mathbf{u}' = \mathbf{A}^{\dagger}\mathbf{r}^{I}, \tag{4.80}$$

and the estimation of b and $\bar{\mathbf{f}}$ now equal to

$$\tilde{b} = u'(1), \tag{4.81}$$

$$\tilde{\bar{\mathbf{f}}} = [u'(2), u'(3), \cdots, u'(L+1)]^T, \tag{4.82}$$

respectively, where $u'(i)$ is the i-th element of vector $\mathbf{u}'$.

4.4 Conclusions

For wideband systems, the frequency dependency of IQ imbalance cannot be ignored. In this chapter, we discussed the frequency dependent IQ imbalance estimation and compensation. We first looked at the TX only IQ imbalance estimation and compensation, and then looked at the joint RX IQ imbalance and frequency offset estimation and compensation. Li et al. [25] proposed a method to do joint frequency dependent TX and RX IQ imbalance estimation and compensation in the frequency domain. To investigate the time domain method that can deal with both TX and RX IQ imbalances and also can estimate and compensate frequency offset simultaneously is an interesting research topic.

References

1. "IEEE 802.11ad standard draft D0.1," [Available] http://www.ieee802.org/11/Reports/tgadupdate.htm.
2. M. Dohler, R. W. Heath, A. Lozano, C. B. Papadias, and R. A. Valenzuela, "Is the PHY layer dead?," *IEEE Commun. Mag.*, pp. 159–165, Apr. 2011.
3. M. Valkama, M. Renfors, and V. Koivunen, "Compensation of frequency-selective I/Q imbalances in wideband receivers: Models and algorithms," *Proc. SPAWC 2001*, Taoyuan, Taiwan, R.O.C., March 20–23, 2001
4. Y. Tsai, C.-P. Yen, and X. Wang, "Blind frequency-dependent I/Q imbalance compensation for direct-conversion receivers," *IEEE Trans. Wireless Commun.*, vol. 9, no. 6, pp. 1976–1986, Jun. 2010.
5. L. Anttila, M. Valkama, and M. Renfors, "Circularity-based I/Q imbalance compensation in wideband direct-conversion receivers," *IEEE Trans. on Veh. Technol.*, vol. 57, no. 4, pp. 2099–2113, Jul. 2008.
6. L. Anttila, M. Valkama, and M. Renfors, "Frequency-selective I/Q mismatch calibration of wideband direct-conversion transmitters," *IEEE Trans. on Circuits and Systems-II: Express Briefs.*, vol. 55, no. 4, pp. 359–363, Apr. 2008.

7. O. Mylläri, L. Anttila, and M. Valkama, "Digital transmitter I/Q imbalance calibration: real-time prototype implementation and performance measurement," *18th European Signal Processing Conference (EUSIPCO-2010)*, Aalborg, Demark, Aug. 23–27, 2010.
8. G. Xing, M. Shen, and H. Liu, "Frequency offset and I/Q imbalance compensation for direct-conversion receivers," *IEEE Trans. Wireless Commun.*, vol. 4, no. 2, pp. 673–680, Mar. 2005.
9. H. Lin, X. Zhu, and K. Yamashita, "Low-complexity pilot-aided compensation for carrier frequency offset and I/Q imbalance," *IEEE Trans. Commun.*, vol. 58, no. 2, pp. 448–452, Feb. 2010.
10. Y.-C. Pan and S.-M. Phoon, "A time-domain joint estimation algorithm for CFO and I/Q imbalance in wideband direct-converstion receivers," *IEEE Trans. Wireless Commun.*, vol. 7, no. 11, pp. 2353–2361, Nov. 2012.
11. A. Schuchert and R. Hasholzner, "A novel IQ imbalance compensation scheme for the receiption of OFDM signals," *IEEE Trans. Consum. Electron.*, vol. 47, no. 3, pp. 313–318, Aug. 2001.
12. K. P. Pun, J. E. Franca, C. Azeredo-Leme, C. F. Chan and C. S. Choy, "Correction of frequency-dependent I/Q mismatches in quadrature receivers," *Electron. Lett.*, vol. 37, no. 23, pp. 1415–1417, Nov. 2001.
13. S. Simoens, M. de Courville, F. Bourzeix, and P. de Champs, "New I/Q imbalance modeling and compensation in OFDM systems with frequency offset," *Proc. IEEE PIMRC 2002*.
14. H. Minn and D. Munoz, "Pilot designs for channel estimation of MIMO OFDM systems with frequency-depedent I/Q imbalances," *IEEE Trans. Commun.*, vol. 58, no. 8, pp. 2252–2264, Aug. 2010.
15. B. Narasimhan, S. Narayanan, H. Minn, and N. Al-Dhahir, "Reduced-complexity baseband compensation of joint Tx/Rx I/Q imbalance in mobile MIMO-OFDM," *IEEE Trans. Wireless Commun.* vol. 9, no. 5, pp. 1720–1728, May 2010.
16. B. Narasimhan, D. Wang, S. Narayanan, H. Minn, and N. Al-Dhahir, "Digital compensation of frequency-dependent joint Tx/Rx I/Q imbalance in OFDM systems under high mobility," *IEEE J. Sel. Topics in Signal Process.* vol. 3, no. 3, pp. 405–417, Jun. 2009.
17. M. Marey, M. Samir, and O. A. Dobre, "EM-based joint channel estimation and IQ imbalances for OFDM systems," *IEEE Trans. Broadcast.*, vol. 58, no. 1, pp. 106–113, Mar. 2012.
18. J. Feigin and D. Brady, "Joint transmitter/receiver I/Q imbalance compensation for direct conversion OFDM in packet-switched multipath environments," *IEEE Trans. Signal Process.*, vol. 57, no. 11. pp. 4588–4593, Nov. 2009.
19. T. Schenk, P. Smulders, and E. Fledderus, "Estimation and compensation of TX and RX IQ imbalance in OFDM-based MIMO systems," *Proc. IEEE Radio and Wireless Symposium*, pp. 215–218, 2006.
20. R. Chrabieh and S. Soliman, "IQ imbalance mitigation via unbiased training sequences," *Proc. IEEE Globecom 2007*.
21. E. Lopez-Estraviz, S. D. Rore, F. Horlin, A. Bourdoux, and L. Van der Perre, "Pilot design for joint channel and frequency-dependent transmit/receive IQ imbalance estimation and compensation in OFDM-based transceivers," *Proc. IEEE ICC*, 2007.
22. E. Lopez-Estraviz and L. Van der Perre, "EM based frequency-dependent transmit/receive IQ imbalance estimation and compensation in OFDM-based transceivers," *Proc. IEEE Globecom*, 2007.
23. J. Luo, W. Keusgen, and A. Kortke, "Preamble designs for efficient joint channel and frequency-selective I/Q-imbalance compensation in MIMO-OFDM systems," *Proc. IEEE WCNC 2010*.
24. Y. Zhou and Z. Pan, "Impact of LPF mismatch on I/Q imbalance in direct conversion receivers," *IEEE Trans. Wireless Commun.*, vol. 10. no. 6, pp. 1702–1708, Jun. 2011.
25. Y. Li, L. Fan, H. Lin, and M. Zhao, "Simple Method to Separate and Simultaneously Estimate Channel and Frequency Dependent TX/RX IQ Imbalances for OFDM Systems," submitted, Mar.2013.